Collins

Let's Practise 2

Maths Workbook

Revised Edition

Preface

***Let's Practise* (Revised Edition)** is a series of eight workbooks for classes 1 to 8, which are aimed at giving comprehensive practice to support child's development of mathematical skills. This series is planned primarily as a series of worksheets presented in a variety of simple and engaging styles. The worksheets are benchmarked against the guidelines of the *National Curriculum Framework* and NCERT syllabus requirements.

As practice is of the utmost importance in mathematics for strengthening the child's foundation in this subject, care has been taken to help develop essential understanding and key maths skills in students. The worksheets are planned in such a way that they become progressively more challenging to ensure complete understanding and application of the concepts.

This series has been completely revised to facilitate comprehensive learning, with a large number of questions relating to real-life experience and interests of the students.

The worksheets can be used by teachers for the purpose of assessments, if they want to, as they are designed that way, incorporating a variety of questions.

Key Features

Quick Facts to recall the important points in a chapter

Worksheets in each chapter to reinforce understanding of the fundamental concepts

Revision Worksheet at the end of each chapter to provide focused and comprehensive practice

Mental Maths to develop essential mental maths skills

Challenge to sharpen thinking skills

Puzzle Corner to enhance the application of the concepts in a fun way

Value-based Question to understand the basic values through the application of practical aspects

Reasoning Worksheets to help develop reasoning and problem-solving skills

Though we have done our best to incorporate feedbacks from our users, suggestions and comments from all are most welcome.

Authors

Contents

1. Shapes 7
2. Counting in Groups 10
3. Light and Heavy 15
4. Counting in Tens 18
5. Patterns 22
6. More on Shapes 27
7. Capacity 30
8. Tens and Ones 34
9. Time: Days and Months 38
10. Additions of 2-digit Numbers 42
11. Lines and Its Types 46
12. Addition and Subtraction 50
13. Measurement 54
14. More on Addition and Subtraction 56
15. Data Handling 61
Reasoning Worksheets 65
Answers 69

1 Shapes

Quick Facts

1. The objects with straight edges or flat surfaces slide.
2. The objects with no edges or curved surfaces roll.
3. The objects with both flat and curved surfaces roll as well as slide.

WORKSHEET 1
Long and Round Objects

1. Description of some objects is given below. Guess their names.

a. Flat at both ends but round like a straw ....

b. Has no corner and is round in appearance ....

c. Flat at its 4 corners with straight edges ....

d. Has a tip at one end but is curved at the other end ....

e. Curved at both ends but is long in appearance ....

2. Name and classify the following objects that are (a) round and (b) not round in shape.

3. Name and draw any three things that are (a) round in shape (b) long in appearance.

WORKSHEET 2

Things that Either Slide, Roll or Both

1. **Write R for the objects that roll and S for the objects that slide.**

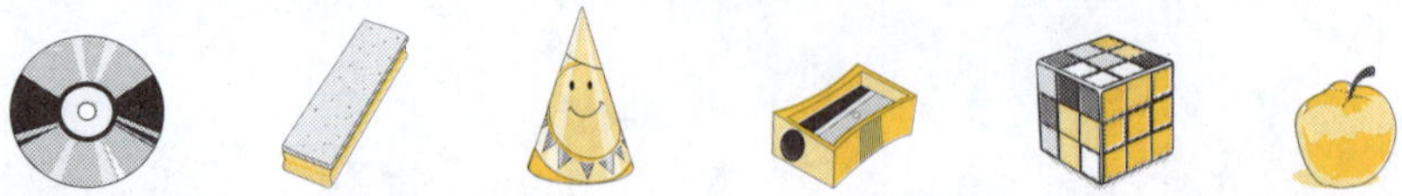

2. **Encircle the objects that can both roll and slide.**

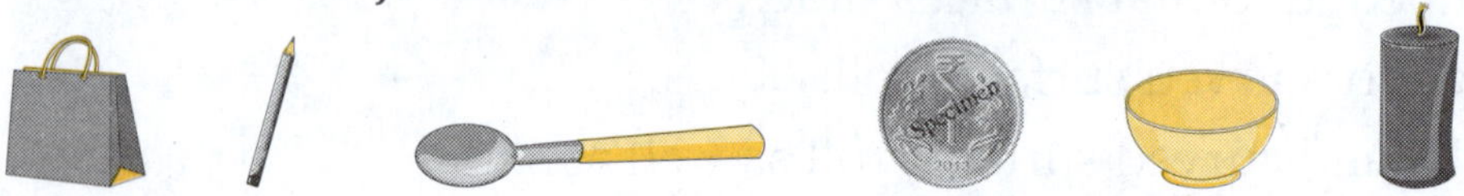

3. **Look at your surroundings. Draw four things that can**

   a. only roll   b. only slide   c. both roll and slide

Challenge

I have 4 corners and 4 edges. All my sides are equal. Who am I? Also, draw me.

Revision Worksheet

1. **Fill in the blanks using the words given below.**

   roll   long   slide   round

   a. A candle is ....   b. An egg will ....

   c. A bowl is ....   d. A pen will ....

   e. A wheel will ....

2. **Tick (✓) the rolling objects and cross-out (×) the sliding objects.**

   a. Matchbox and onion   b. Lemon and suitcase   c. Cucumber and cube

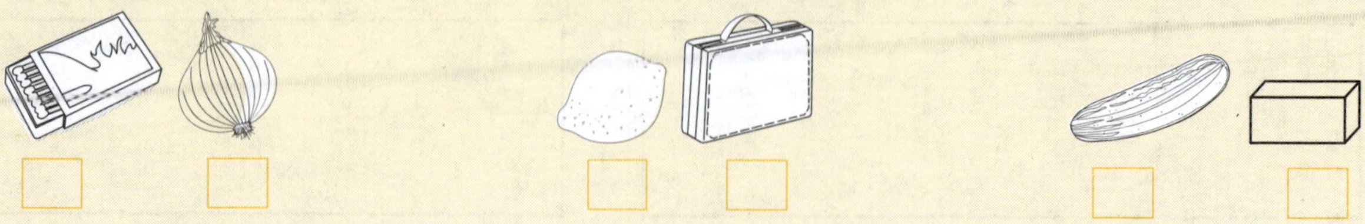

d. Tyre and book

e. Mobile phone and apple

Value-based Question

Meena learnt about different shapes. She asked her teacher why tyres are round in shape and not in some other shape like square or triangle. Write your answer with a reason.

....

Mental Maths

List a thing that

a. is long b. is round c. rolls d. slides e. rolls and slides

Puzzle Corner

The names of a few things are hidden in the word search puzzle given below. Find them and classify them as long or round.

L	K	Q	T	M	E	S	N	N	E
L	J	L	L	C	P	T	X	X	G
R	I	W	T	X	O	I	T	P	N
N	O	C	M	A	R	C	X	Z	A
B	W	B	N	O	X	K	N	D	R
U	J	W	E	E	T	D	D	U	O
N	T	G	Q	N	P	A	L	M	F
H	W	R	V	Y	I	E	M	T	T
K	Z	K	I	Q	R	O	R	O	X
L	A	J	N	I	R	B	C	K	T

Things that are round: ....

Things that are long: ....

2 Counting in Groups

Quick Facts

1. Ordinal numbers define the position of things.
2. Numbers can be counted by grouping.
3. The given set of numbers can be arranged either in ascending order or in descending order.
4. Arranging numbers from the smallest to the biggest is ascending order and arranging numbers from the biggest to the smallest is descending order.
5. Numbers are compared using lesser than (<), greater than (>) or equal to (=) symbol.

WORKSHEET 1
Ordinal Numbers

1. Match the ordinal numbers with the corresponding numerals.

a. 5	i. Ninth
b. 8	ii. Sixth
c. 2	iii. Fifth
d. 9	iv. Eighth
e. 6	v. Second

2. Identify the ordinal numbers from the jumbled letters.

a. IFSRT b. DIRHT c. OUTFHR d. NEVSTEH e. TNEHT

3. Identify the ordinal number using the clues given.

a. I stand between second and fourth. Who am I? ....

b. I stand in front of second. Who am I? ....

c. I stand just after ninth. Who am I? ....

d. I stand between fifth and seventh. Who am I? ....

e. I stand just after tenth. Who am I? ....

4. **The final round of a 500-m running race was going on. At some point, Kiran was behind Madhu. Amit was at the last while Bhanu ran in between Veena and Amit. Mark their names on the track given below accordingly.**

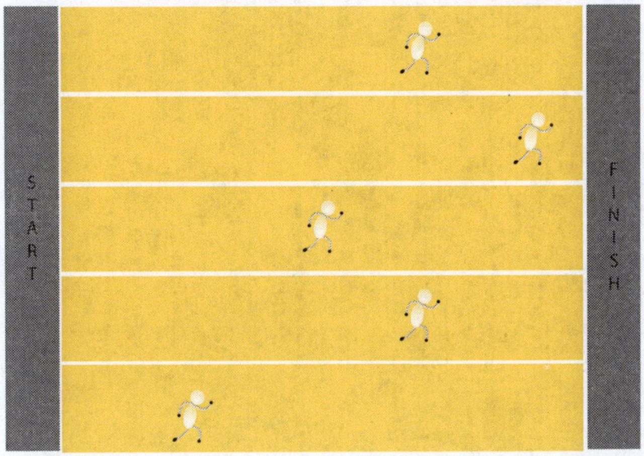

WORKSHEET 2
Counting in Groups

1. **Count the number of objects in groups and match with the corresponding number.**

a.  i. 28

b. 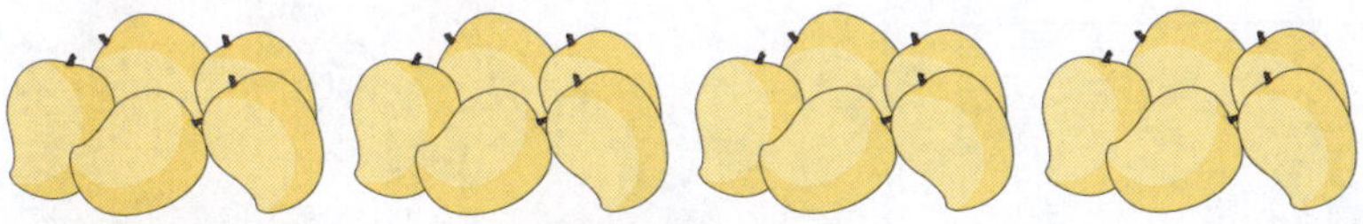 ii. 18

c.  iii. 15

d.  iv. 20

e.  v. 14

2. **Count and write the total number in each of the following.**

a.  .... tomatoes

b.  .... mangoes

c.  .... leaves

d.  .... flowers

e.  .... balls

3. **Complete the following table.**

	Number of groups	Number of objects in each group	Total number of objects
a.	5	4	______
b.	8	______	24
c.	______	9	27
d.	2	10	______
e.	______	7	28

4. **Write all the possible ways of grouping 24 balls.**

....

WORKSHEET 3
Comparison of Numbers

1. **Fill in the missing numbers in each of the following.**

a. 65, 66, _____, _____, 69 b. 20, _____, 22, 23, _____ c. 33, 34, 35, _____, _____

d. 87, _____, _____, 90, 91 e. _____, _____, 54, 55, 56

2. **Compare the following numbers using < or > or =.**

a. 63 ____ 36 b. 48 ____ 84 c. 65 ____ 65 d. 53 ____ 85 e. 90 ____ 95

3. **Arrange the following numbers in ascending order.**

a. 6, 75, 43, 38, 52 b. 80, 8, 38, 27, 98 c. 76, 34, 46, 57, 75

d. 64, 53, 51, 49, 47 e. 39, 82, 73, 46, 51

4. **Arrange the following numbers in descending order.**

a. 7, 85, 67, 38, 45 b. 65, 84, 28, 91, 43 c. 37, 48, 56, 87, 12

d. 68, 63, 72, 43, 19 e. 13, 85, 42, 67, 90

Revision Worksheet

1. **Count the total number of objects in each of the following and write it in the space provided.**

a.	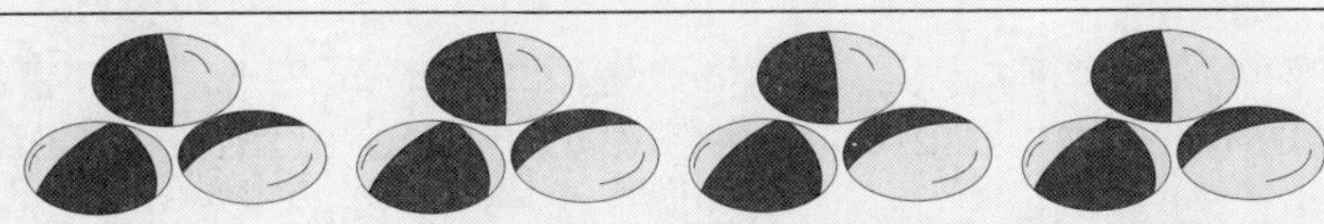 = ____ balls
b.	 = ____ petals
c.	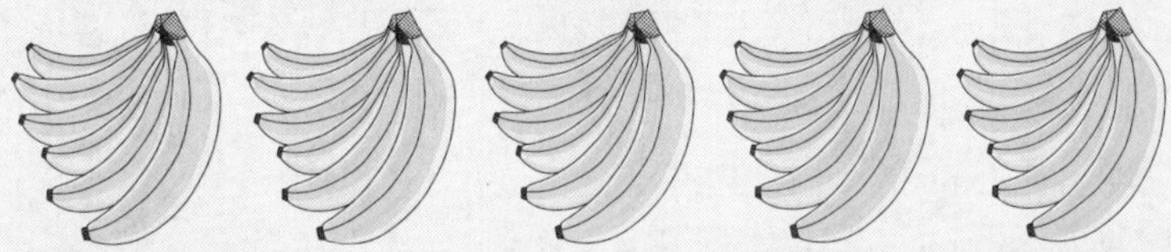 = ____ bananas
d.	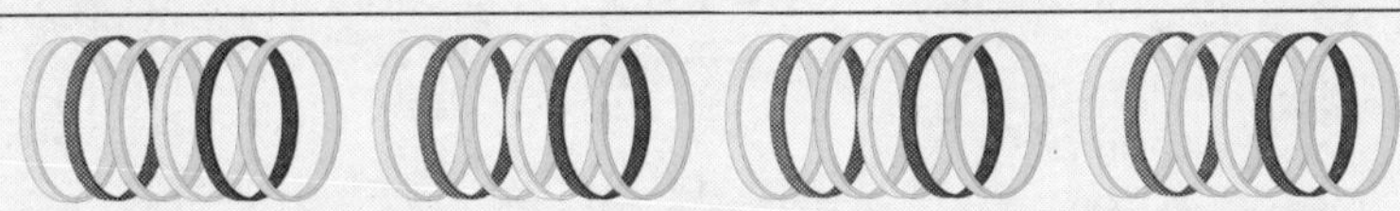 = ____ bangles
e.	 = ____ legs

2. **State whether the following statements are true (T) or false (F).**

a. If 4 oranges are kept in a group and there are 5 such groups, the total number of oranges will be 20. ☐

b. The position between eighth and tenth is eleventh. ☐

c. 56 is more than 65. ☐

d. 79 is in between 78 and 80. ☐

e. The numbers 85, 84, 83, 82 are in ascending order. ☐

3. **Write the position of the following months present in a year.**

a. March is the .... month of a year.

b. July is the .... month of a year.

c. September is the .... month of a year.

d. February is the .... month of a year.

e. May is the .... month of a year.

4. Identify the order of numbers and write 'Ascending/Descending' in the space given.

a. 56, 48, 39, 24, 19 ....

b. 34, 68, 84, 90, 92 ....

c. 96, 83, 60, 49, 19 ....

d. 85, 61, 57, 34, 28 ....

e. 28, 43, 51, 68, 83 ....

Challenge

Form all possible 2-digit numbers using digits 7, 5, 3 and 0 and arrange them in the descending order.
(*Note*: Digits should not be repeated while forming the numbers.)

Puzzle Corner

Find the ordinal numbers from first to tenth in the grid and encircle them. Write the words and the numbers corresponding to them in the blanks.

m	t	s	n	k	i	s	d	r	h
a	s	e	c	o	n	d	t	t	t
b	e	o	h	f	s	r	n	j	h
d	v	u	t	o	f	i	r	s	g
l	e	m	f	u	n	h	x	o	i
k	n	f	i	r	s	t	q	t	e
t	t	s	f	t	e	n	t	h	h
w	h	r	t	h	g	i	f	t	i

....
....
....

3 Light and Heavy

Quick Facts

1. An object with more weight is heavier than an object with less weight.
2. In a balance, the pan having the object with less weight will be higher and the pan having the object with more weight will be lower.

WORKSHEET
Heavier to Lighter

1. Tick (✓) the heavier ones and cross-out (×) the lighter ones in each section.

a. i.  ☐	ii.  ☐
b. i.  ☐	ii.  ☐
c. i.  ☐	ii.  ☐
d. i.  ☐	ii.  ☐

2. Write (H) for the heavier object and (L) for the lighter object.

a. i. Carrot  ☐ ii. Bunch of bananas 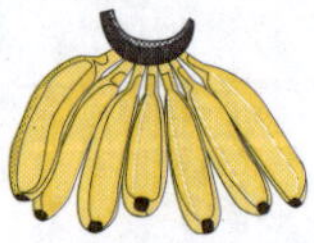 ☐	b. i. Pencil  ☐ ii. Book 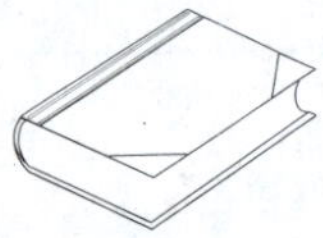 ☐
c. i. Football  ☐ ii. Tennis ball  ☐	d. i. Hammer  ☐ ii. Nail  ☐

3. Fill in the blanks using the words *'heavier'* or *'lighter'*.

a. An elephant is .... than a rabbit.
b. A pen is .... than a book.
c. A hen is .... than its egg.
d. A mobile phone is .... than a laptop.
e. A squirrel is .... than a rabbit.

4. Choose the correct option.

a. Which is the heaviest vegetable among the following?
   i. Pumpkin ii. Onion iii. Garlic iv. Beans
b. Which is the lightest object among the following?
   i. Pen ii. Pencil iii. Book iv. Bag
c. Which is the heaviest animal among the following?
   i. Fox ii. Dog iii. Tiger iv. Rabbit
d. Which is the heaviest vehicle among the following?
   i. Scooter ii. Tractor iii. Car iv. Bicycle
e. Which is the lightest object among the following?
   i. Cell phone ii. Laptop iii. Television iv. Refrigerator

Puzzle Corner

Colour the smileys to make the balance equal.

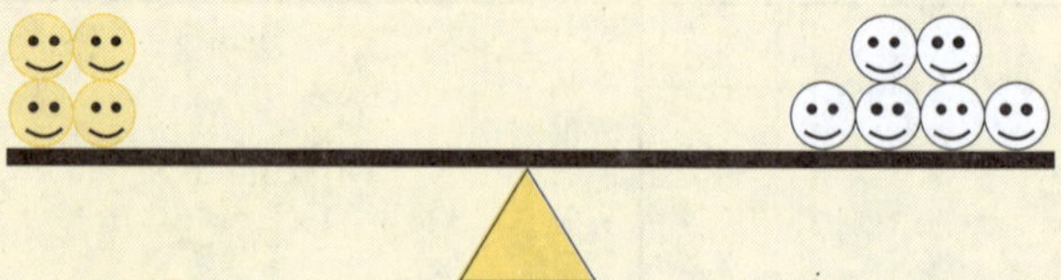

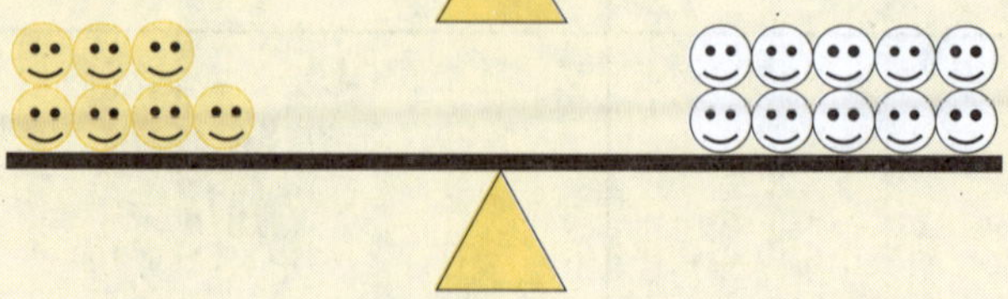

Revision Worksheet

1. Encircle the animal that is heavier in each of the following.

a.	b.	c.	d.

2. Write (L) for lighter and (H) for heavier in the boxes provided.

a.	b.	c.	d.
			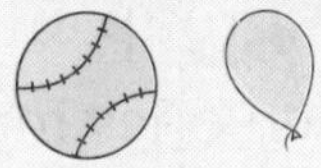
i. ☐ ii. ☐	i. ☐ ii. ☐	i. ☐ ii. ☐	i. ☐ ii. ☐

3. Match the following according to their weight by drawing lines.

A	a. 	b. 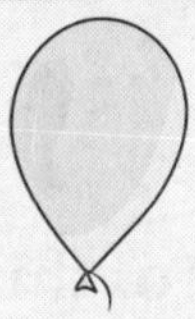	c. 	d. 	e. 
B	i. 	ii. 	iii. 	iv. 	v. 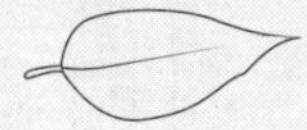

Challenge

If Tom is heavier than Jerry and lighter than Spike, who is the heaviest and who is the lightest among the three?

4 Counting in Tens

Quick Facts

1. Numbers can be represented by grouping them into groups of tens.
2. Counting becomes easier when numbers are grouped in tens.
3. A group of ten items can be arranged in different patterns.
4. Ten ones make 1 ten.

WORKSHEET
Counting in Tens

1. Count in groups of ten and write the total number of things in each of the following.

a. 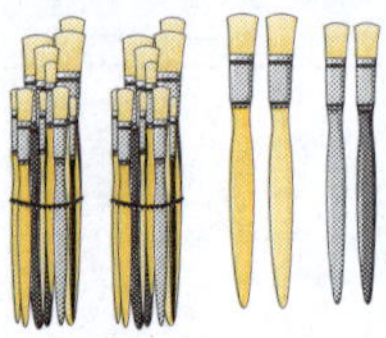

b. 

c. 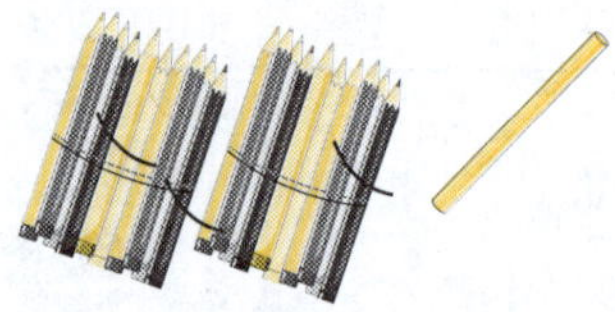

d. 

e. 

2. Fill in the blanks.

a. 10 + 6 = ...   b. 30 + ... = 34   c. .... + 2 = 72

d. 65 = 60 + ...   e. 80 + 9 = ...

3. **Group the given objects in tens corresponding to the given number. One has been done for you.**

16 

a. 30 

b. 48 

c. 65 

d. 52

e. 77 

4. **A balloon seller stood in front of a birthday hall. He counted the number of balloons with him.**

Answer the following questions.

a. How many bunches of 10 balloons are there? ....

b. How many balloons are there altogether? ....

After selling some balloons, he counted them again.

c. There are .... bunches of 10 balloons left.

d. There are a total of .... + 2 = ... balloons in all.

e. .... – .... = ... balloons were sold by him.

Challenge

List all the combinations in tens that make a sum of 50.

....

Revision Worksheet

1. Match the number with its representation.

a. 64	i. 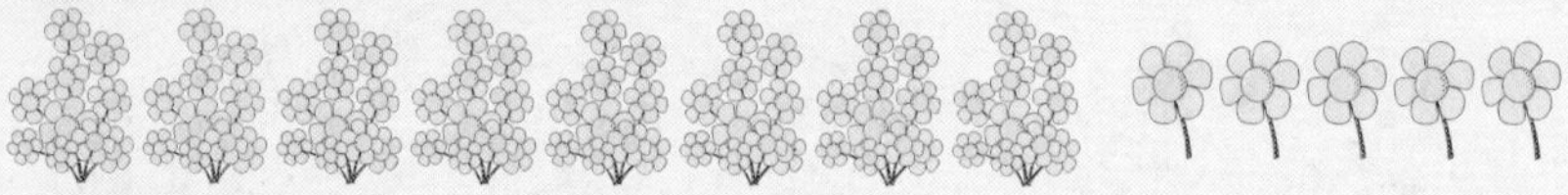
b. 85	ii. 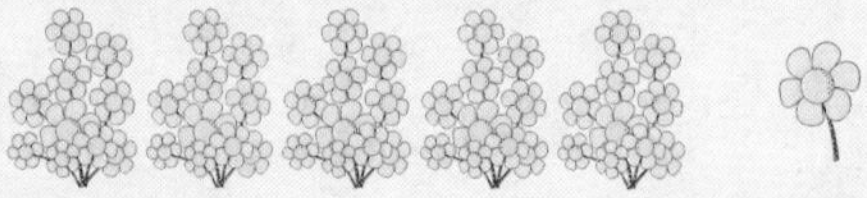
c. 72	iii. 
d. 51	iv. 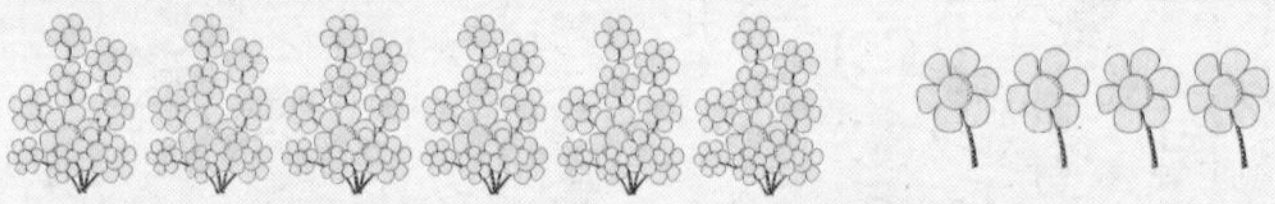
e. 36	v. 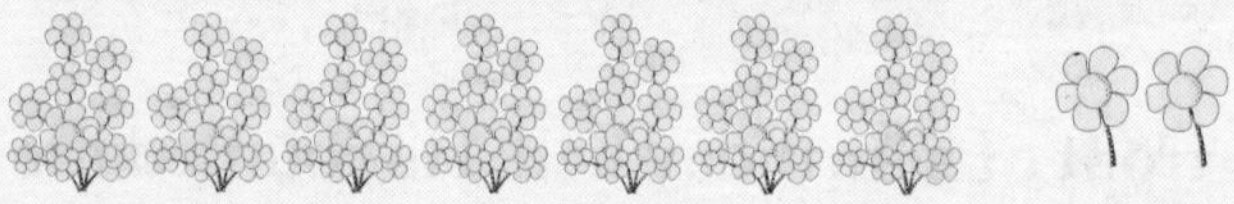

2. Draw different patterns to represent tens using any shape you like.

3. My mother placed some pencils on the table as shown below. Look at the picture and answer the questions that follow.

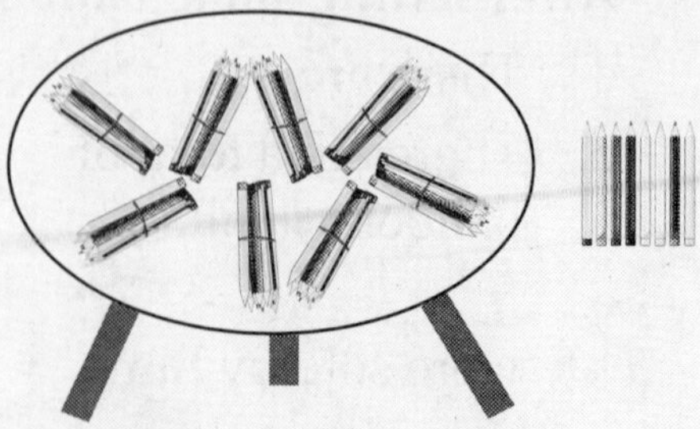

a. How many bundles of 10 pencils are there on the table? ....

b. How many pencils are not placed on the table? ....

c. What is the total number of pencils? ....

d. Write the mathematical statement to represent all the pencils. ....

Value-based Question

Vani distributed notebooks among 10 children in an orphanage. She gave each child a bag having 10 notebooks. How many notebooks did she distribute in all? What value can we learn from Vani?

....

Puzzle Corner

Complete the following circle by counting in tens.

Mental Maths

a. .... + ... = 25

b. .... + 9 = 99

c. 42 = 40 + ...

d. 6 packets of 10 chocolates each make .... chocolates.

e. 3 + 5 + 2 = ...

5 Patterns

Quick Facts

1. If shapes, numbers or letters are repeated in a particular manner, they form patterns.
2. Number patterns can be created using numbers with operations like addition, subtraction, multiplication and division.

WORKSHEET 1
Shape Patterns

1. Extend the pattern by colouring the following objects.

a. 

b. 

c. 

d. 

2. Extend the given shape patterns.

a. 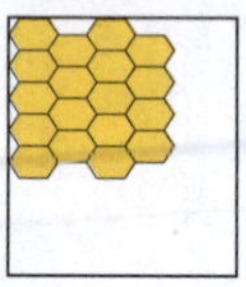

b. 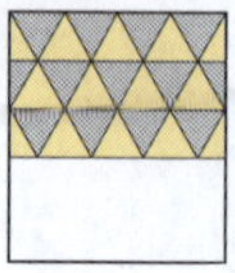

c. 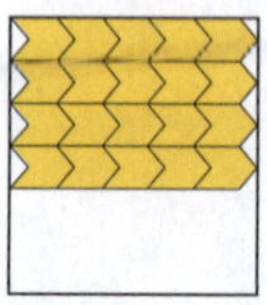

d. 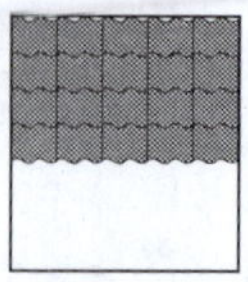

3. **Encircle the shape that can be used to create a floor pattern without leaving any gap.**

a. 

b. 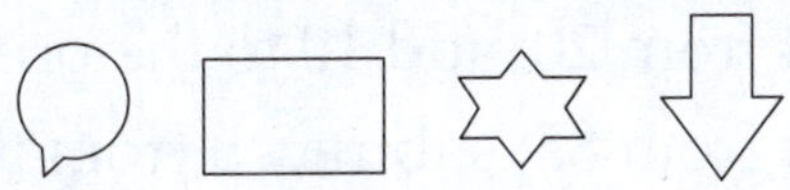

c. 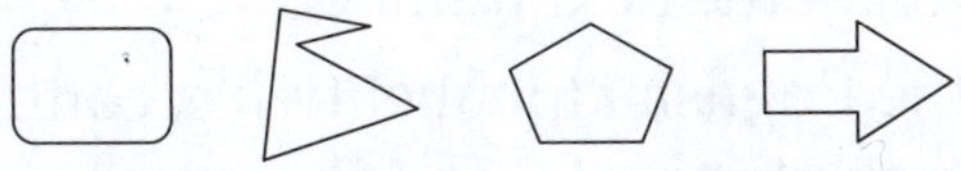

d. 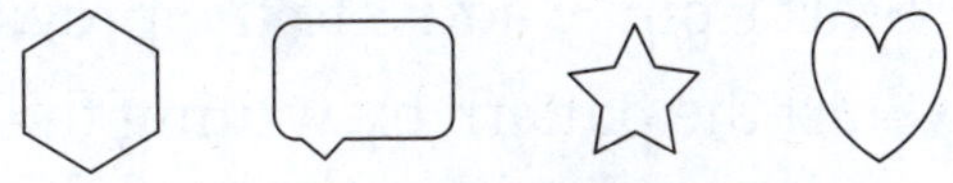

4. **Identify the mistake in the following patterns.**

a. 

b. 

c. 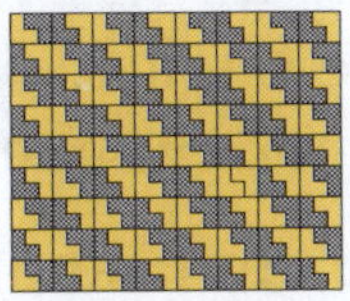

d. 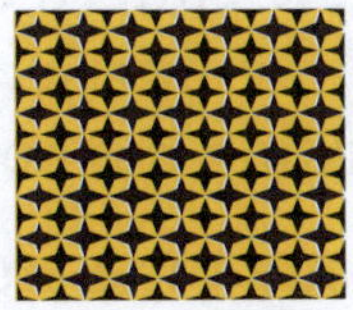

5. **Create your own floor pattern using different geometrical shapes and mention the rule followed to create the pattern.**

WORKSHEET 2

Number Patterns

1. **Identify the rule followed in each pattern.**

	Pattern	Rule followed
a.	15, 20, 25, 30, 35	....
b.	17, 19, 21, 23, 25	....
c.	90, 80, 70, 60, 50	....
d.	2, 4, 8, 16, 32	....
e.	4, 10, 16, 22, 28	....

2. **Extend the given patterns.**

a. AB, BC, CD, DE, EF, ...

b. 1, 12, 123, 1234, 12345, ...

c. Right, Left, Left, Right, Left, ...

d. A1, B2, C3, D4, E5, ...

e. AAA, BBB, CCC, DDD, EEE, ...

3. **Complete the following patterns.**

a. 7, 12, 17, ..., 27

b. ...., 89, 79, 69, 59

c. 5, ..., 9, 11, 13

d. AZ, BY, ..., DW, EV

e. MN, NM, MN, ..., MN, NM

4. **Create a pattern using the given rules. Write five entries for each of the following.**
   a. Start from 20; add 10 to the previous number to get the next number.
   b. Start from 85; subtract 5 from the previous number to get the next number.
   c. Start from 6; add 3 to the previous number to get the next number.
   d. Start the pattern by writing the last letter of the English alphabet twice; continue the pattern by writing the previous letter, twice each time.
   e. Start the pattern by writing the first letter of the English alphabet and number 1 and again the first letter of the English alphabet after the number; continue the pattern by writing the next letter and number 2 and again the same letter after the number.
5. **Create your own number pattern and mention the rule followed in creating the pattern.**

Revision Worksheet

1. **Identify the pattern and write the rule in the blank provided.**

	Pattern	Rule followed
a.	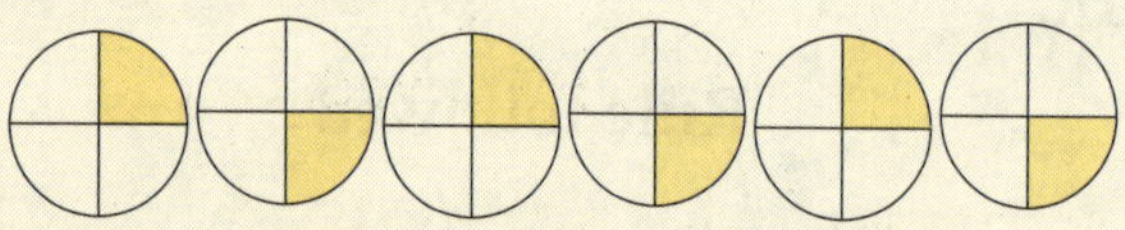	....
b.		....
c.	AA1, BB2, CC3, DD4, EE5, FF6	....
d.	6, 10, 14, 18, 22, 26	....

2. **Extend the following patterns.**

   a.  ....

b. 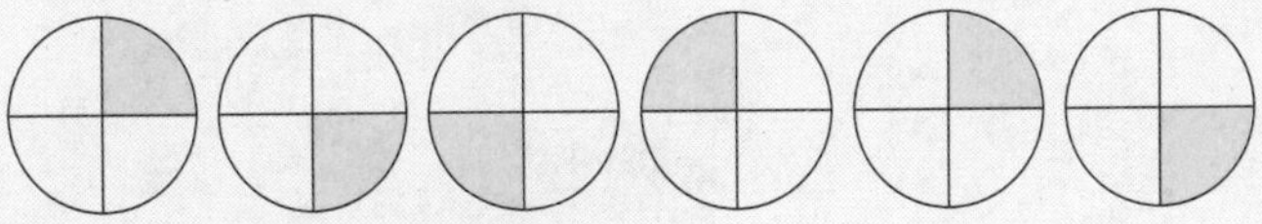 ....

c. 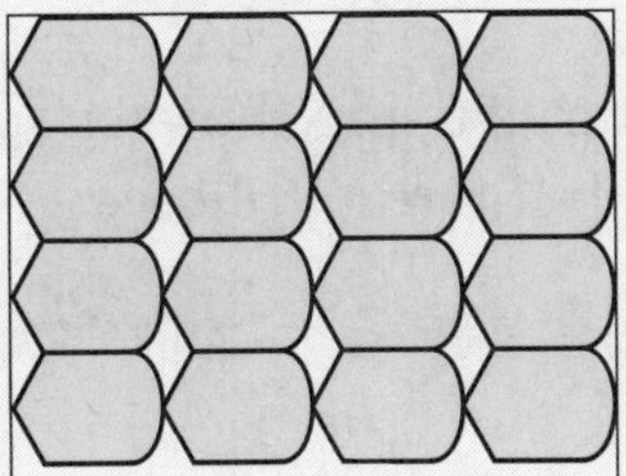

....

d. 7, 14, 21, 28, 35, 42 ....

3. Complete the patterns.

a. 81, 76, 71, ..., 61

b. 8, 16, 24, ..., 40

c. A1, AA2, ..., AAAA4, AAAAA5

d. 101, 1001, ..., 100001, 1000001

4. Identify the mistake in the following patterns.

a.	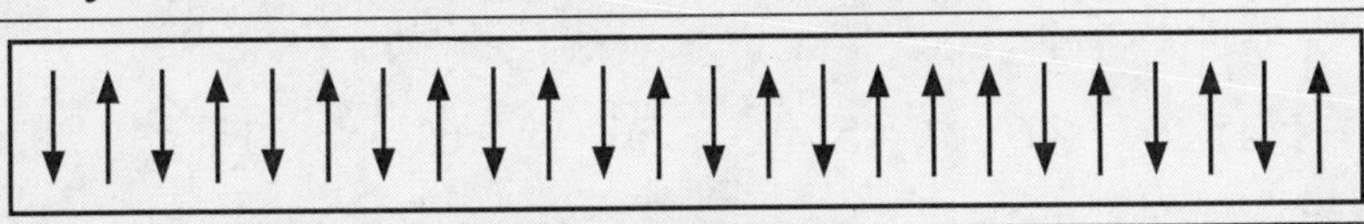
b.	
c.	
d.	ABC, BCD, CDE, DEF, EFE

5. Create patterns by following the instructions given below.

a.	Starts with ; it is followed by  and 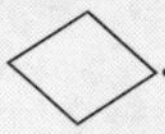. The same is repeated further.
b.	Four  continuously and four  continuously. The pattern with both these shapes continues by reducing one each time.
c.	Start the pattern with number 5 and add 3 to the previous number to get the next number.
d.	Start the pattern with the first number followed by the first alphabet and the same number again. Continue with the next number and next alphabet.

Challenge

There are 3 circles, 3 triangles and 6 rectangles. Create patterns using these shapes only once. Draw the possible combination of patterns along with the rule followed to create the pattern.

Puzzle Corner

The following grid contains 5 number patterns consisting of 6 numbers each. The patterns are in horizontal, vertical and slanting order. Find and encircle them.

17	45	23	13	23	43	53	63
23	36	41	46	51	56	61	14
19	44	33	53	55	58	63	68
33	52	51	52	54	55	56	57
13	60	53	49	58	63	78	88
18	68	43	50	57	64	57	50
79	76	67	64	61	58	55	52

6 More on Shapes

Quick Facts

1. Different objects can be traced to obtain geometrical shapes like oval, circle, rectangle and triangle.
2. Circle, triangle, square and rectangle are the basic geometrical shapes.

WORKSHEET
Footprints and shapes

1. Match the animal with its footprint.

a. 	b. 	c. 	d. 
i. 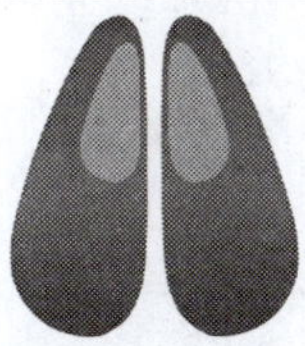	ii. 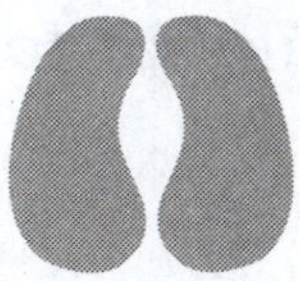	iii. 	iv. 

2. Write the names of the following shapes.

Shapes	Names	Shapes	Names
a. 		c. 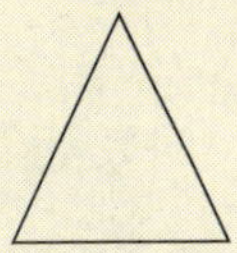	
b. 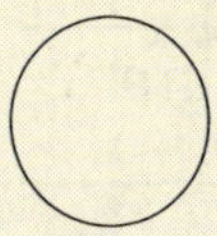		d. 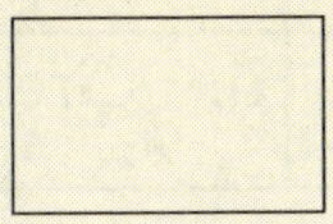	

3. **Colour the rectangles blue, circles brown, triangles green and squares red. Also, find the number of each shape present in the given picture and write them in the space provided.**

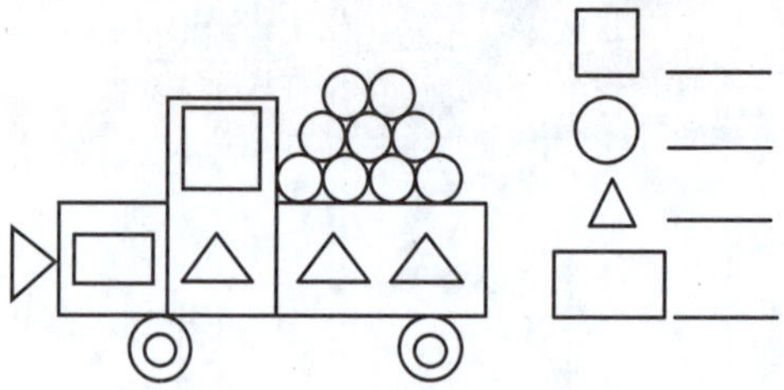

4. **Trace the following objects and write the name of the shape represented by each object.**

Object	Trace below	Name of the shape
a. Eraser		
b. Chessboard		
c. Coin		

Revision Worksheet

1. **Write the names of the objects that look like the following shapes.**
   a. Circle: ....
   b. Square: ....
   c. Rectangle: ....
   d. Triangle: ....
2. **Match the following.**

a. 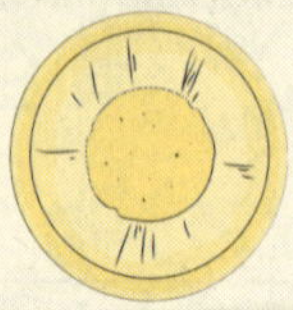	b. 	c. 	d. 
i. Triangle	ii. Circle	iii. Rectangle	iv. Square

3. Using all the four shapes (circles, squares, triangles and rectangles), draw your own picture in the space given below.

Challenge

Encircle the pumpkin that has square-shaped eyes, triangle-shaped nose and circle-shaped mouth.

Puzzle Corner

Look at the picture and fill in the blanks.

My hat has:

..... circles, ... rectangles, ... squares, ... triangles

My face has:

..... circles, ... rectangles, ... squares, ... triangles

My neck is a .... **. My arms are** .... **. My body is a** .... **. My legs are** .... **.**

Each of my feet looks like a .... **and is split into 2** ....

7 Capacity

Quick Facts

1. Different containers have different capacities.
2. Smaller containers hold lesser amounts and bigger containers hold more amounts.
3. Smaller containers can be used for pouring and measuring.
4. Capacities of different containers can be compared.
5. We use different amounts of water for doing different things.

WORKSHEET 1
Estimation of Capacity

1. A shop sold cold *badam* milk in summer. There was a  bucket full of *badam* milk. They sold it in two different glasses.

Tall glass  ₹25

 Small glass ₹10

a. Which is easier to pour into one tall  glass: a bucket or a jug?

b. Sania found that 4 tall glasses filled a jug.
Vidya wants to buy one jug of *badam* milk. How much does she need to pay?

c. Manu found that three short glasses filled one  tall glass.

How many  short glasses of *badam* milk will fill the  jug?

d. How many short glasses will fill half the jug?

e. How much should Kiran pay, if she buys 8 short glasses of *badam* milk?

2. Guess how many  **will fill the following containers.**

a.  b. 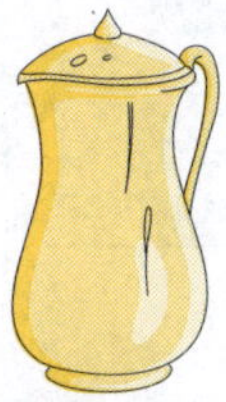 c.  d. 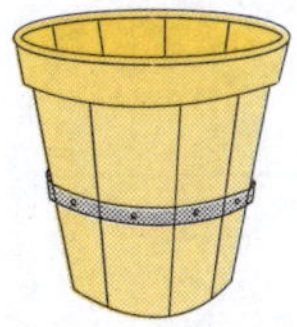 e. 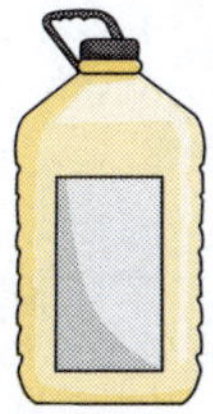

3. Find out how much water is used by your family every day for the following activities. Also, name the container used.

Activity	Amount of water
Taking bath	
Washing clothes	
Cooking and cleaning utensils	
Drinking	

WORKSHEET 2
Comparison of Capacities

1. Encircle the container that holds the least amount of water.

a. 
i. ii.

b. 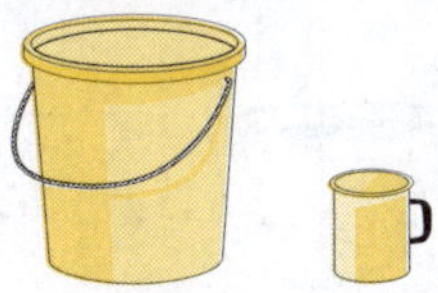
i. ii.

c. 
i. ii.

d. 
i. ii.

e. 
i. ii.

2. Tick the container that holds the most amount of water.

a. 
i. ii. iii

b. 
i. ii. iii

c. 
i. ii. iii

3. What are the different containers used in your kitchen? Draw any 5 of them.

4. Consider this bowl  .

Write what will happen if water from the bowl is poured into each of the following containers. Will they be completely filled or water will pour out?

a.  b.  c. 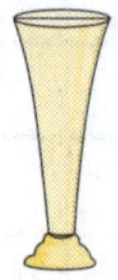 d. 

....

....

Puzzle Corner

Arrange the following in decreasing order of their capacities to hold water.

Challenge

In which of the following figures are the mugs arranged in ascending order?

a. 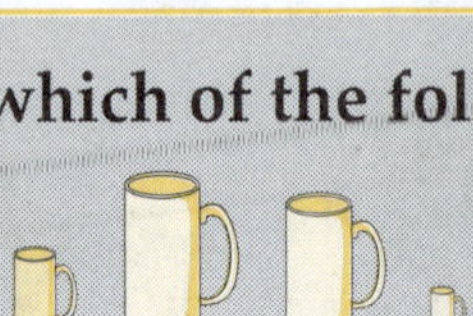 b.  c. 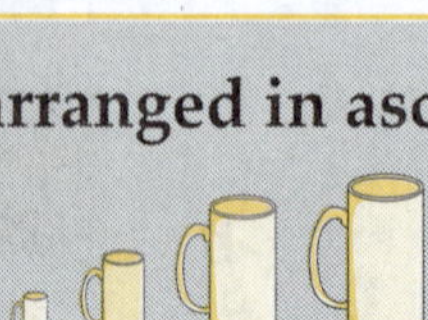 d. 

Mental Maths

a. Which will hold more, ink filler or spoon?
b. How many drops of milk can a spoon hold?
c. Name a container that can hold large amounts of water.
d. How many buckets of water would be needed to wash a vehicle?

Revision Worksheet

1. Name any 3 containers that hold less water than a jug.

....

2. Name any 3 containers that hold more water than a mug.

....

3. There was a big  jug of sugarcane juice. A lady was selling a  glass of it for ₹15.

a. If Banu bought 5 glasses of sugarcane juice, then how much did she pay?
b. Manu found that 12 glasses filled the jug.
Anil and his friends drank sugarcane juice from half of the jug. How much did they pay?
c. Guess how many jugs of sugarcane juice will fill this container. 
d. How many glasses of sugarcane juice can you drink?

Value-based Question

Buttermilk was prepared in buckets. They were served to the guests at a function. Some old people who were thirsty stopped in front of the house. Banu gave the buttermilk to them also. Which container did Banu use for serving all of them? What value do you learn from this?

....

8 Tens and Ones

Quick Facts

1. There are different denominations of notes and coins. Rupee is used in India and its symbol is ₹.
2. Coins are available in the denominations of 1, 2, 5 and 10 rupees.
3. Notes are available in the denominations of 5, 10, 20, 50 rupees, etc.
4. Ten ones make one ten.

WORKSHEET 1
Tens and Ones

1. Match the amount with the possible combinations of money.

a. ₹12

b. ₹9

c. ₹25

d. ₹18

e. ₹6

i. 

ii. 

iii. 

iv. 

v. 

2. **Fill in the blanks.**

a. 37 = ... + 7

b. 59 = 50 + ...

c. 78 = ... + 8

d. 65 = 60 + ...

e. 99 = ... + 9

3. **Write the number representing the following tens and ones.**

a. 4 tens 3 ones = ...

b. 6 tens and 9 ones = ...

c. 9 tens and 6 ones = ...

d. 3 tens and 8 ones = ...

e. 9 tens and 0 ones = ...

4. **Encircle the shapes given below in groups of tens and write the number represented by them in the right column. One has been done for you.**

a. 	b. 	c. 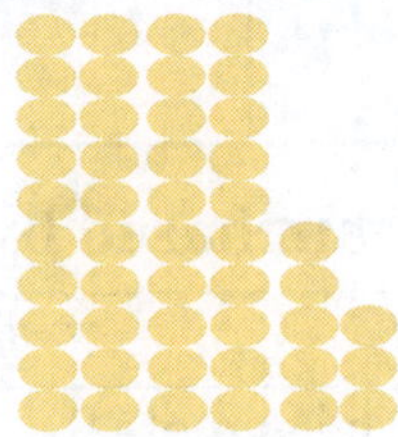	d. 
Number 35			

Challenge

Write any 8 different combinations of notes and coins that you can make for ₹40.

WORKSHEET 2
Real Life Applications

1. **Write at least two different combinations of money to buy these items.**

a. 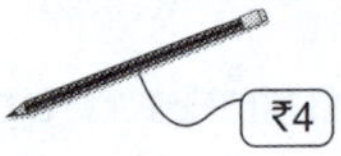	b. 	c. 	d. 	e. 

2. **Different types of sweets purchased by a school from a sweet stall are as given below.**

Sweets	*Gulab jamun*	*Burfi*	*Jalebi*
Number of sweets	43	37	19

Answer the following.

a. If *gulab jamuns* are to be distributed among 4 groups of 10 students each, how many *gulab jamuns* will be left after distribution?

b. If 7 *barfis* are left after distributing the sweets to Class 1 students, find the number of students in Class 1.

c. In Class 2, there are 20 students. If *jalebis* are distributed to each student of Class 2, how many students will not get *jalebis*?

d. How many sweets were bought altogether by the school?

e. If there are 90 students in the school, find the number of sweets left over after distributing all the sweets to the students.

3. **School students are being taken on a picnic in vans. The capacity of each van is 10 students. The number of students in the classes is given in the table below. Look at the table and answer the questions that follow:**

Class 1	Class 2	Class 3	Class 4	Class 5	Total
20	34	36	19	26	135

a. How many vans will be needed for all the students? ....

b. How many vans are required for Class 5 students? ....

c. If there are only 7 vans in the school, how many more vans must be booked for the picnic? ....

d. For Class 2, if only 3 vans are allotted, how many students will be left over?.............

e. How many vans are required for Classes 2 and 3 together? ....

Revision Worksheet

1. Write the given numbers in terms of tens and ones. One has been done for you.

a. 98 = 9 tens + 8 ones b. 55 = ...

c. 27 = ... d. 81 = ...

e. 99 = ...

2. Tick (✓) the true statements and cross (×) the false statements.

a. ₹5 + ₹2 + ₹1 + ₹1 = ₹8 

b. ₹2 + ₹2 +₹2 + ₹2 + ₹2 = ₹10 

c. ₹10 + ₹10 + ₹5 + ₹2 + ₹1 = ₹29

d. ₹10 + ₹10 + ₹10 + ₹10 +₹1 + ₹2 = ₹43

e. ₹20 + ₹10 + ₹10 + ₹5 + ₹2 = ₹47

3. Manu and Anu are playing bingo card games. If  represents 10 points, and  represents 1 point, answer the following.

a. Manu won  cards and  cards. So, he scored a total of .... points.

b. Anu won  cards. So she scored .... points.

c. Represent the points scored by them together:

....

d. How many points were scored by them together? .... tens .... ones.

9 Time: Days and Months

Quick Facts

1. There are seven days in a week: Monday, Tuesday, Wednesday, Thursday, Friday, Saturday and Sunday.
2. There are 12 months in a year: January, February, March, April, May, June, July, August, September, October, November and December.

WORKSHEET 1
Days of the Week

1. **Write the days for the following.**
   a. Third day of the week: ....
   b. First day of the week: ....
   c. Fourth day of the week: ....
   d. Seventh day of the week: ....
   e. Fifth day of the week: ....

2. **Complete the following time table by looking at your class timetable.**

Period	Art and craft	English	GK	Music	Games
Day of the Week					

3. **The table given below shows the names of the students who will present the 'Thought of the Day' at the assembly.**

Day	**Name of the Student**
Monday	Varun
Tuesday	Malini
Wednesday	Tanu
Thursday	Dhruv
Friday	Soumya
Saturday	Balu

State whether the following are true (T) or false (F).

a. Soumya's turn is on the day after Friday. ☐
b. Varun's turn is on the day before Tuesday. ☐
c. Balu's turn is on the day after Sunday. ☐
d. Malini's turn is on the day before Wednesday. ☐
e. Dhruv's turn is on the day after Thursday. ☐

WORKSHEET 2
Months in a Year

1. Write the names of your friends and their birthday month.

Name of the friend	Birthday month

2. Write the months in the blanks provided.

a. .... comes before April.

b. November comes after ....

c. .... is the last month of the year.

d. .... comes after June.

e. January comes before ....

3. Do as directed.

a. List the months having 30 days.

....

b. List the months having 31 days.

....

c. How many days are there in June and July together? ....

d. How many Saturdays are there in October this year? ....

e. How many days are there in February this year? ....

4. Write the month corresponding to each festival by looking at the calendar.

a. Pongal: ....

b. Ramzan: ....

c. Christmas: ....

d. Gandhi Jayanti: ....

e. Raksha Bandhan: ....

5. Complete the following table.

My favourite	Name	Name of the month when it grows
Flower		
Fruit		
Vegetable		

Revision Worksheet

1. **Fill in the blanks.**
   a. If the first day of the week is Sunday, then the 5th day of the week is ....
   b. The name of the month that starts with the letter D and has 31 days in it is ....
   c. The name of the month that has only 3 letters in it is ....
   d. The 10th month of the year is ....
   e. The number of days in the month of November is ....

2. **Match the following.**

a.	Number of days in a week	i.	31
b.	Number of days together in March and August	ii.	30
c.	Number of days in the month of February	iii.	7
d.	Number of days in the month of June	iv.	62
e.	Number of days in the month of December	v.	28

3. **If today is Thursday, then which day is**
   a. tomorrow? ....
   b. 3 days before it? ....
   c. 2 days after it? ....
   d. 7 days from it? ....
   e. 4 days before it? ....

Mental Maths

1. Which day comes between Sunday and Tuesday?
2. Name the fourth month of the year.
3. .... is the sixth day of the week.
4. Ninth month of the year is ....
5. August comes after ....
6. Children's day is celebrated in the month of ....

Challenge

1. Name the month that has the most number of letters.
2. Which day was the day before yesterday, if the day after tomorrow is a Thursday?

Value-based Question

Dhruv celebrates his birthday in the fifth month of the year, on the third day of the week along with the people of an old age home. He donates a part of his earnings to buy essential things for them. Name the day and month in which he celebrates his birthday.

What value do you learn from Dhruv?

....

Puzzle Corner

Find the names of the days and the months by solving the hints given below.

F B J A N U A R Y
Y E Y A D S E U T
A J B B L B Q Y X
D U N R M M A Y Y
I N B E U D J B T
R E C T N A R W M
F E T U W R R T D
D N S G G Z Z Y L

1. The month that has the least number of days ....
2. The month that has the least number of letters ....
3. A day after Monday ....
4. Two days before Sunday ....
5. 6th month of the year ....
6. The last day of the week ....
7. First month of the year ....
8. Last month of the year ....

10 Addition of 2-digit Numbers

Quick Facts

1. Counting together is addition.
2. The '+' symbol denotes addition and it is read as 'plus'.
3. The number we get after addition is called sum or total and the numbers added are called addends.

WORKSHEET 1
Addition of 2 Numbers

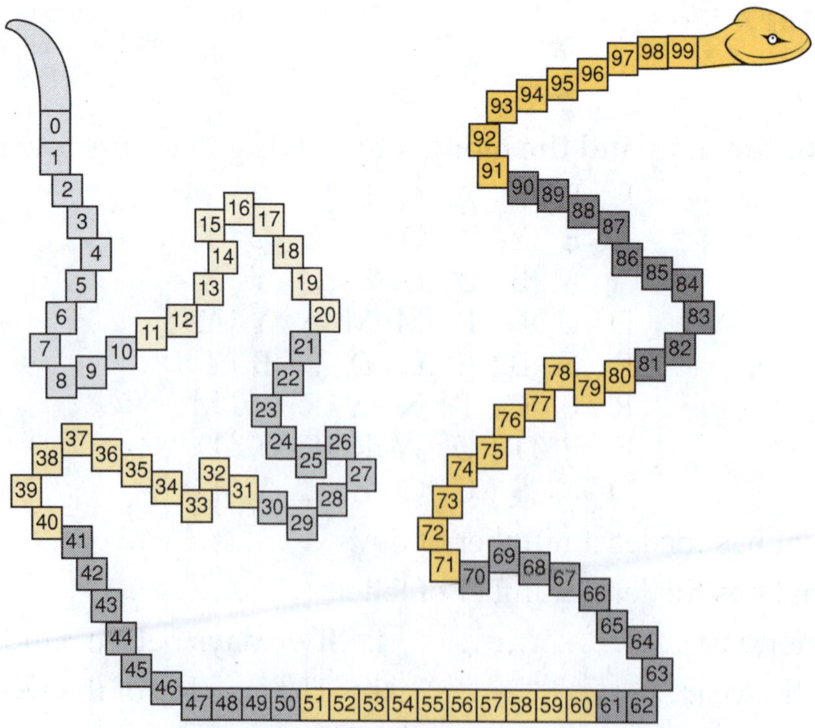

Note: Use the number chain for forward counting.

1. **Add the following numbers.**

a. 23 + 38 = ... b. 19 + 16 = ... c. 17 + 47 = ...

d. 54 + 39 = ... e. 56 + 35 = ...

2. **Find the missing numbers.**

a. 37 + ... = 42 b. 56 + ... = 63 c. 75 + ... = 81

d. 19 + ... = 23 e. 21 + ... = 30

3. **Encircle the addend for the given sum.**

	Sum	Addends		
a.	33	15 + 18	15 + 17	15 + 16
b.	43	20 + 20	20 + 23	20 + 33
c.	54	25 + 24	25 + 25	25 + 29
d.	65	30 + 35	30 + 30	25 + 55
e.	74	29 + 44	24 + 45	29 + 45

4. **Match the following.**

a.	35 + 49	i.	91
b.	26 + 28	ii.	72
c.	19 + 64	iii.	84
d.	34 + 57	iv.	54
e.	29 + 43	v.	83

WORKSHEET 2
Addition of 3 or More Numbers

1. **Add the following numbers.**

a. 7 + 8 + 9 = ... b. 4 + 9 + 3 = ... c. 8 + 5 + 4 + 9 = ...

d. 6 + 1 + 8 + 7 = ... e. 3 + 4 + 5 + 6 + 7 = ...

2. **Write the possible addends for the following.**

a. 18 = ... + ... + ...

b. 27 = ... + ... + ...

c. 33 = ... + ... + ... + ...

d. 20 = ... + ... + ... + ...

e. 15 = ... + ... + ... + ...

3. **Encircle the addend for the given sum.**

	Sum	Addend		
a.	29	12 + 13 + 4	12 + 13 + 15	12 + 13 + 16
b.	15	10 + 3 + 1	10 + 3 + 2	10 + 3 + 3
c.	52	40 + 10 + 2	40 + 10 + 3	40 + 10 + 4
d.	38	30 + 5 + 1	30 + 5 + 2	30 + 5 + 3
e.	10	5 + 5 + 5	5 + 3 + 2	5 + 3 + 1

4. **State whether the following are true (T) or false (F).**

a. 9 + 6 + 7 = 22 ☐

b. 5 + 4 + 3 = 12 ☐

c. 9 + 8 + 7 + 1 = 24 ☐

d. 8 + 4 + 6 + 7 = 25 ☐

e. 8 + 8 + 7 + 9 = 31 ☐

Challenge

Observe the first number tree and complete the next one.

Puzzle Corner

Fill in the grid with numbers such that their addition is equal to the sum given horizontally, vertically or diagonally.

				17
2		5		22
6	4	0		16
	9			19
	2		1	7
9	23	15	17	14

Revision Worksheet

1. Some of the places in the number grid are occupied by birds and animals. Write at least 2 possible addends for each number occupied by them.

1	2	3	4	5	6	7	8	9	10
11	12	13	14	15	16		18	19	20
21	22	23	24	25	26	27	28	29	30
31	32		34	35	36	37	38	39	40
41	42	43	44	45	46	47	48	49	50
51	52	53	54	55		57	58	59	60
61		63	64	65	66	67	68	69	70
71	72	73	74	75	76	77	78	79	80
81	82	83	84	85	86	87		89	90
91	92	93	94	95	96	97	98	99	100

a.  b.  c.  d.  e.

2. Find the missing numbers.

a. 25 + ... = 30 b. 17 + ... = 25

c. .... + 10 = 19 d. .... + 12 = 15

e. 23 + 22 = ...

3. Match the following.

a.	10 + 12	i.	36
b.	10 + 12 + 13	ii.	22
c.	10 + 12 + 14	iii.	24
d.	10 + 14 + 13	iv.	35
e.	10 + 14	v.	37

4. State whether the following statements are true (T) or false (F).

a. 13 + 10 = 22 ☐ b. 12 + 10 + 10 = 32 ☐

c. 10 + 25 + 20 = 45 ☐ d. 10 + 15 = 25 ☐

e. 10 + 15 + 10 = 35 ☐

11 Lines and Its Types

Quick Facts

1. Lines can be straight or curved.
2. Straight lines can be sleeping, standing or slanting.
3. Letters of the English alphabet and numbers can be written using lines.
4. Lines can be used for drawing patterns and designs.

WORKSHEET 1
Types of Lines

1. Write if each of the following is a straight or a curved line.

a.

....

b. 

....

c. 

....

d. 

....

e.

....

f. 

....

2. Count the number of curved lines and straight lines in each of the the given figure.

a. 
Number of straight lines: ....
Number of curved lines: ....

b. 
Number of straight lines: ....
Number of curved lines: ....

c. 
Number of straight lines: ....
Number of curved lines: ....

3. **Classify the following letters into the given 3 categories based on the type of lines used to write them.**

A B C D E F G H I J K L M N O P Q R S T U V W X Y Z

Straight lines	Curved lines	Both straight and curved lines

WORKSHEET 2

Shapes and Pictures with Lines

1. **Encircle the object made of straight lines in each pair.**

a.  i. 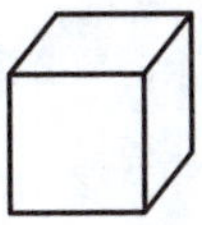 ii.

b.  i.  ii.

c. 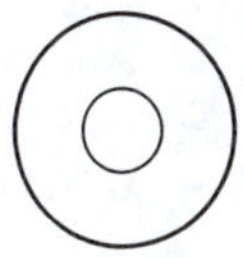 i.  ii

d. 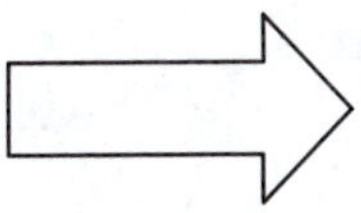 i.  ii.

e.  i. 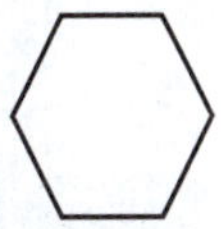 ii.

2. **Write the number of standing, sleeping, slanting and curved lines used in the following figures.**

Figure	Standing lines	Sleeping lines	Slanting lines	Curved lines
a. 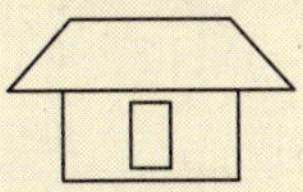				
b. 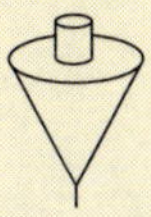				
c. 				
d. 				

3. **Classify the shapes given below into the following 3 categories based on the types of lines used to draw them.**

Only straight lines	Only curved lines	Both straight and curved lines

4. **Draw as instructed.**

a. A figure using only straight lines.	b. A figure using only curved lines.	c. A figure using both straight and curved lines.

Revision Worksheet

1. **Draw the following lines.**

a. Standing line	b. Slanting line
c. Sleeping line	d. Curved line

2. **Identify the lines in the following figures and write them in the given box (straight lines, curved lines or both straight and curved lines). One has been done for you.**

a.  Both straight and curved lines

b.  c. 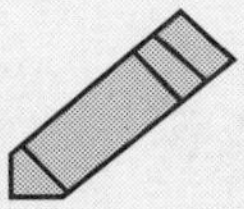 d. 

3. Observe the given figure and put a tick (✓) for the correct statements and a cross (×) for incorrect statements.

a. The house has only straight lines.

b. The tree has only curved lines.

c. The sun has both straight and curved lines.

d. The house has same shapes.

4. Classify the following letters and numbers based on the type of lines used to write them.

C O R N F L A K E S 1 2 3 4 5 6 7 8 9 0

Straight lines	Curved lines	Both straight and curved lines

Challenge

Join the alphabets to complete the picture. Count and write the number of straight and curved lines used to join the alphabets.

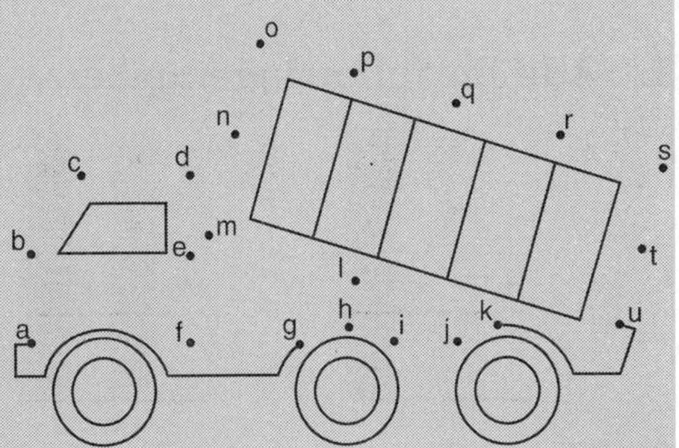

Puzzle Corner

Draw a human face using curved lines.

12 Addition and Subtraction

Quick Facts

1. Counting things together is called addition.
2. The symbol of addition is '+'.
3. Add the numbers in the ones place first and then in the tens place.
4. The symbol for subtraction is '–'.
5. Subtraction means taking away some objects from the given objects.

WORKSHEET 1
Addition

1. Add the following numbers.

a.

	T	O
	3	3
+		7

b.

	T	O
	4	5
+	1	5

c.

	T	O
	3	6
+	2	4

d.

	T	O
	6	7
+	1	8

e.

	T	O
	5	4
+	2	9

2. Circle the mistake and write the correct answer.

a.

	T	O
	4	2
+		8
	4	0

	T	O
	4	2
+		8

b.

	T	O
	5	2
+	1	8
	6	1

	T	O
	5	2
+	1	8

3. **Colour the two numbers whose sum is equal to the given number. One has been done for you.**

60	61	9	51	8

a. 40 — 15, 25, 10, 35

b. 35 — 32, 17, 22, 18

c. 44 — 6, 18, 26, 24

d. 94 — 65, 55, 19, 39

WORKSHEET 2
Subtraction

1. **Subtract the following numbers.**

a.

	T	O
	2	5
−		7

b.

	T	O
	1	4
−		8

c.

	T	O
	3	1
−		9

d.

	T	O
	4	3
−		5

e.

	T	O
	5	5
−		6

2. a.

	T	O
	3	5
−	1	8

b.

	T	O
	5	4
−	3	6

c.

	T	O
	6	1
−	2	9

d.

	T	O
	7	3
−	4	5

e.

	T	O
	9	2
−	5	8

3. **Subtract the numbers given below and colour the circle with the correct answer.**

a. 26 – 8 = (18) (19) (17)

b. 32 – 13 = (18) (19) (17)

c. 43 – 6 = (37) (38) (39)

d. 75 – 26 = (47) (48) (49)

e. 89 – 59 = (40) (20) (30)

f. 99 – 36 = (63) (53) (73)

Puzzle Corner

Add vertically and horizontally to fill in the missing numbers.

8	+	5	+	10	=	
+		+		+		+
9	+	7	+	10	=	
+		+		+		+
10	+	10	+	9	=	
=		=		=		=
	+		+		=	

WORKSHEET 3
Word Problems

Solve the following.

1. In a class, there are 27 girls and 18 boys. How many students are there in the class?
2. Vani had ₹43 and Dhruv had ₹39. How much money did they have altogether?
3. Anusha's necklace had 35 beads and Banu's necklace had 29 beads. Find the total number of beads in both the necklaces.
4. There were 98 seats in the theatre; 67 were occupied by the audience. How many seats were left unoccupied?
5. A milk vendor sells 75 litres of milk every month. His cow gives 90 litres of milk. He uses the remaining milk for himself. How much milk does he use for himself?

Challenge

Observe the sequence of addition given in Tree A and complete the missing numbers in Tree B.

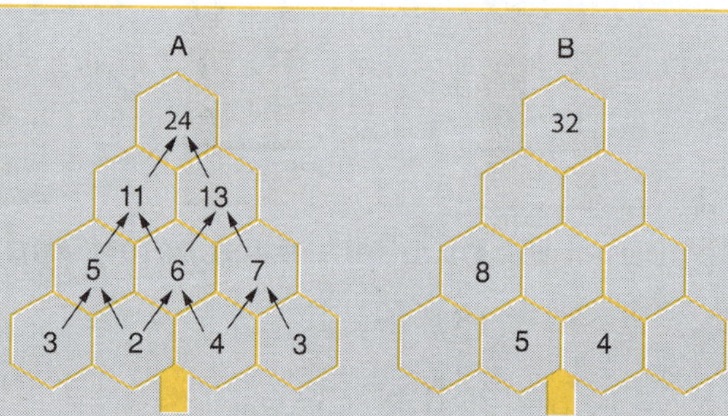

Value-based Question

Ram and Shyam are best friends. Shyam brought a number puzzle but he is unable to understand it. He asks Ram to help him. Ram looks at the puzzle and tells Shyam that he has to fill in the missing numbers so that the sum on both sides of the '=' sign is equal. Now, solve this puzzle and check if you get the same answers as Shyam gets.

a. 13 + ☐ = 11 + 4 b. 9 + 8 = ☐ + 14 c. 10 + ☐ = 12 + 3 d. 15 + 2 = ☐ + 7

e. ☐ + 14 = 12 + 5 f. ☐ + 12 = 9 + 8 g. 2 + 14 = ☐ + 13 h. ☐ + 12 = 14 + 4

Revision Worksheet

1. Match the following.

a. 16 + 34 b. 61 – 49 c. 79 + 11 d. 55 – 38 e. 99 – 79

i. 12 ii. 20 iii. 50 iv. 17 v. 90

2. Fill in the blanks.

a. The sum of 64 and 27 is …….
b. The difference between 55 and 32 is …….
c. The total of 41 and 19 is …….
d. ……. is 35 more than 50.
e. 42 less than 71 is …….

3. Find the following.

a. $\begin{array}{r} 4\ 3 \\ +\ 2\ 9 \\ \hline \\ \hline \end{array}$ b. $\begin{array}{r} 3\ 4 \\ -\ 1\ 3 \\ \hline \\ \hline \end{array}$ c. $\begin{array}{r} 5\ 9 \\ +\ 3\ 6 \\ \hline \\ \hline \end{array}$ d. $\begin{array}{r} 8\ 0 \\ -\ 4\ 8 \\ \hline \\ \hline \end{array}$ e. $\begin{array}{r} 9\ 3 \\ -\ 6\ 5 \\ \hline \\ \hline \end{array}$

4. Solve the following word problems.

a. Out of 90 mangoes, 55 mangoes were used for preparing milkshake. How many mangoes were left unused?
b. Tara had 58 stickers and Sonali had 36 stickers. How many stickers do they have in total?
c. Team A scored 44 points during the first round of the quiz and 35 points during the second round. What is the total score of Team A?
d. A bus ticket costs ₹66. You paid ₹80 for it. How much will you get back?
e. Manav had 72 coloured papers, out of which 36 were green and the rest were red. How many red-coloured papers did he have?

Mental Maths

1. 15 + 25 = **2.** 63 + 8 = **3.** 57 + 4 = **4.** 72 + 8 = **5.** 55 + 5 =

6. What number should be added to the given number or subtracted from the given number to make the sum equal to the given answer? Find the number and write it in the box.

a. $\begin{array}{r} 6\ 1 \\ +\ \square \\ \hline 7\ 8 \\ \hline \end{array}$ b. $\begin{array}{r} 3\ 4 \\ +\ \square \\ \hline 7\ 8 \\ \hline \end{array}$ c. $\begin{array}{r} 5\ 9 \\ -\ \square \\ \hline 4\ 6 \\ \hline \end{array}$ d. $\begin{array}{r} 6\ 3 \\ -\ \square \\ \hline 4\ 6 \\ \hline \end{array}$ e. $\begin{array}{r} 8\ 8 \\ -\ \square \\ \hline 4\ 6 \\ \hline \end{array}$

13 Measurement

Quick Facts

1. Fingers, handspan, footspan, cubit and pace are some of the non-standard units of measurement.
2. The distance from the tip of your little finger to the tip of your thumb on a stretched hand makes a handspan.
3. The distance from the end of your heel to the tip of your big toe makes a footspan.
4. One pace is equal to one natural step.

WORKSHEET
Finger Length, Handspan, Footspan and Pace

1. Identify the units of measuring length shown below.

a. 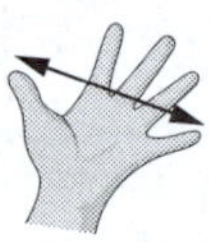

....

b. 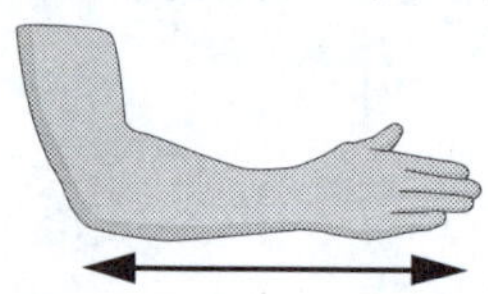

....

c. 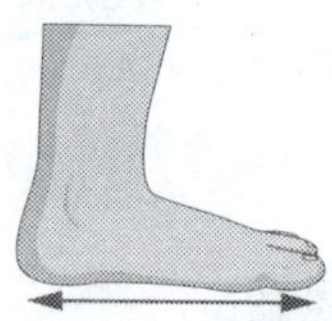

....

d. 

....

2. Choose the correct units of measurement and write them in the boxes given below.

(Handspan or finger length)

The length of a butterfly [ ]

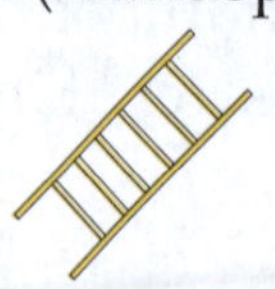

The height of a ladder [ ]

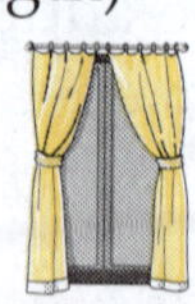

The length of door curtain [ ]

The length of your nose [ ]

3. **Measure each of the following object using paper clips and write them in the space provided.**

a. 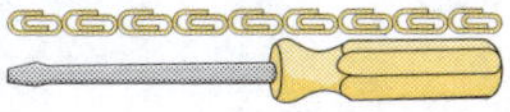

.... Paper clips

b. 

.... Paper clips

c. 

.... Paper clips

d. 

.... Paper clips

Revision Worksheet

1. **Write the length of the given objects.**
   a. Handspan
      Length of the piano = ...
   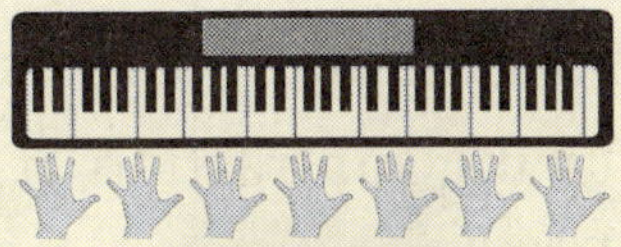
   b. Handspan
      Length of the TV = ...
   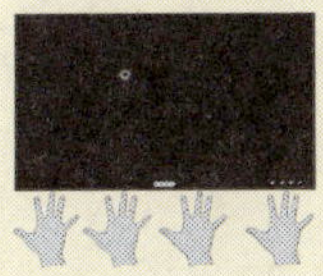
   c. Cubit
      Length of the bamboo = ...
   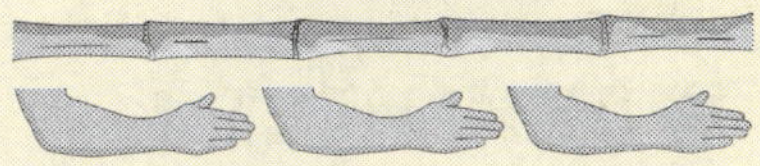
2. **State whether the following statements are true (T) or false (F).**
   a. The distance from the end of your heel to the tip of your big toe is called a handspan. ☐
   b. The distance between the front foot and the back foot makes a footspan. ☐
   c. The length of a handspan is more than the length of a finger. ☐
   d. Footspan is more than the length of a toe. ☐
   e. Cubit is more than the length of a handspan. ☐
3. **List any four objects whose length is lesser than your handspan.**

   ....

Challenge

1. 1 cubit = ... handspans and .... fingers
2. 1 pace = ... footspan
   (*Hint:* Measurement should be made using your own hands and feet.)

14 More on Addition and Subtraction

Quick Facts

1. It is easier to add or subtract by counting in groups of tens.
2. Ten ones are regrouped into one ten and carried over to tens while adding.
3. A smaller number is subtracted from a bigger number.
4. During subtraction, if there are not enough ones to subtract from, regroup one ten to ten ones.

WORKSHEET 1
Addition

1. Add the following numbers.

a. $\begin{array}{r} 23 \\ +\ 18 \\ \hline \\ \hline \end{array}$ b. $\begin{array}{r} 49 \\ +\ 26 \\ \hline \\ \hline \end{array}$ c. $\begin{array}{r} 33 \\ +\ 48 \\ \hline \\ \hline \end{array}$ d. $\begin{array}{r} 54 \\ +\ 37 \\ \hline \\ \hline \end{array}$ e. $\begin{array}{r} 17 \\ +\ 65 \\ \hline \\ \hline \end{array}$

2. Match the following.

a. 22 + 17	i. 72
b. 51 + 29	ii. 39
c. 47 + 25	iii. 93
d. 55 + 38	iv. 91
e. 38 + 53	v. 80

3. Write the numbers of your choice that add up to each of the following.

a. $\begin{array}{r} \square \\ +\ \square \\ \hline 54 \\ \hline \end{array}$ b. $\begin{array}{r} \square \\ +\ \square \\ \hline 83 \\ \hline \end{array}$ c. $\begin{array}{r} \square \\ +\ \square \\ \hline 45 \\ \hline \end{array}$ d. $\begin{array}{r} \square \\ +\ \square \\ \hline 76 \\ \hline \end{array}$ e. $\begin{array}{r} \square \\ +\ \square \\ \hline 91 \\ \hline \end{array}$

4. Solve the following.

a. Datri made 36 garlands in May and 27 garlands in June. How many garlands did she make in all?

b. Varun scored 53 marks in Maths and 46 marks in Science. Find the total marks scored by him.

c. There are 25 bikes and 58 cars in a parking lot. How many vehicles are there altogether?

d. I had ₹64 and my brother gave me ₹16. How much money do I have now?

e. A shop sold 42 cups of coffee and 48 cups of tea. What is the total number of cups of tea and coffee sold by that shop?

WORKSHEET 2
Subtraction

1. Subtract the following.

a. $\begin{array}{r} 3\ 2 \\ -\ 1\ 8 \\ \hline \\ \hline \end{array}$ b. $\begin{array}{r} 5\ 6 \\ -\ 2\ 9 \\ \hline \\ \hline \end{array}$ c. $\begin{array}{r} 7\ 0 \\ -\ 3\ 4 \\ \hline \\ \hline \end{array}$ d. $\begin{array}{r} 9\ 1 \\ -\ 4\ 9 \\ \hline \\ \hline \end{array}$ e. $\begin{array}{r} 8\ 5 \\ -\ 4\ 6 \\ \hline \\ \hline \end{array}$

2. Fill in the blanks.

a. 44 – 28 = ... b. 67 – .... = 39

c. .... – 26 = 47 d. 82 – 25 = ...

e. 71 – 37 = ...

3. Write the numbers of your choice for each of the following.

a. $\begin{array}{r} \square \\ -\ \square \\ \hline 2\ 4 \\ \hline \end{array}$ b. $\begin{array}{r} \square \\ -\ \square \\ \hline 4\ 1 \\ \hline \end{array}$ c. $\begin{array}{r} \square \\ -\ \square \\ \hline 3\ 5 \\ \hline \end{array}$ d. $\begin{array}{r} \square \\ -\ \square \\ \hline 6\ 2 \\ \hline \end{array}$ e. $\begin{array}{r} \square \\ -\ \square \\ \hline 5\ 3 \\ \hline \end{array}$

4. Solve the following.

a. I had 45 pencils. Vani took away 18 pencils from me. How many are remaining with me?

b. Out of 53 chocolates on the table, the children ate 35 of them. Find the number of chocolates left on the table.

c. Manu bought 73 balloons and Anil bought 57 balloons for their birthday celebrations. Who bought more and by how much?

d. Sandhya is 26 years old. Her grandfather is 82 years old. How many years older is her grandfather?

e. How much is ₹46 less than ₹90?

WORKSHEET 3
Real Life Applications

1. Look at the given price list and answer the questions that follow.

Price List

Toys	₹
Aeroplane	55
Bat and Ball	40
Top	22
Rubik's Cube	36
Speaking Parrot	38
Teddy Bear	67
Train	73

a. How much money is needed to buy a rubik's cube and an aeroplane?

....

b. Is ₹65 sufficient to buy a speaking parrot and a top?

....

c. Which two toys can be bought exactly with ₹95?

....

d. Priyan had ₹84. He bought a teddy bear with it. How much money is he left with?

....

e. How much less does a bat and ball cost than a train?

....

2. The children of a particular class were asked to choose a fruit among the following and their choices are tabulated as follows.

Fruit	 Apple	 Watermelon	 Banana	 Mango	 Grapes
Number of students	63	46	56	97	75

Answer the following questions.

a. Children who liked mango are more than those who liked apple. How many more?

....

b. Name the least liked and the most liked fruits.

....

c. How many more children must like banana to reach the same number as grapes?

....

d. How many less children like watermelon than grapes?

....

Revision Worksheet

1. Match the following.

a.
```
   3 5
 + 5 6
 -----
```
b.
```
   4 2
 - 2 5
 -----
```
c.
```
   5 4
 + 3 8
 -----
```
d.
```
   8 3
 - 6 7
 -----
```
e.
```
   7 4
 + 1 6
 -----
```

i. 92   ii. 91   iii. 17   iv. 90   v. 16

2. Tick the facts that are true in the following.

a. 63 – 28 = 35 ☐   b. 42 + 37 = 89 ☐

c. 81 – 48 = 43 ☐   d. 26 + 54 = 80 ☐

e. 95 – 66 = 29 ☐

3. Solve the following word problems.

a. Out of the 95 cookies, 58 of them are distributed. How many cookies are left?

....

b. Srinu had collected 36 stamps. His friend gave him his collection of 45 stamps. How many stamps does he have in all?

....

c. The sum of two numbers is 82. If one number is 49, find the other.

....

d. Anil bought a pen for ₹55 and a notebook for ₹37. How much did he pay the shopkeeper?

....

e. A crayon set costs ₹26 less than a pen costing ₹90. What is the cost of the crayon set?

....

Challenge

Complete the table.

37	+		=	63
+		+		
45	–		=	27
=		=		
	–		=	

Mental Maths

1. 35 + 49 = ...
2. 84 – 15 = ...
3. 99 – 39 = ...
4. 52 + 28 = ...
5. 73 – 35 = ...
6. 34 + 56 = ...
7. .... – 31 = 49
8. 62 – .... = 45

Value-based Question

An old man was wandering in a street. He was asking for food from people. Dhruv bought a packet of *idli* for ₹35 and gave him. He also gave him a water bottle and biscuits spending ₹48 on it. How much did he spend altogether? What value do you learn from this act of Dhruv?

....

....

Puzzle Corner

Complete the pyramid of numbers according to the given operation.

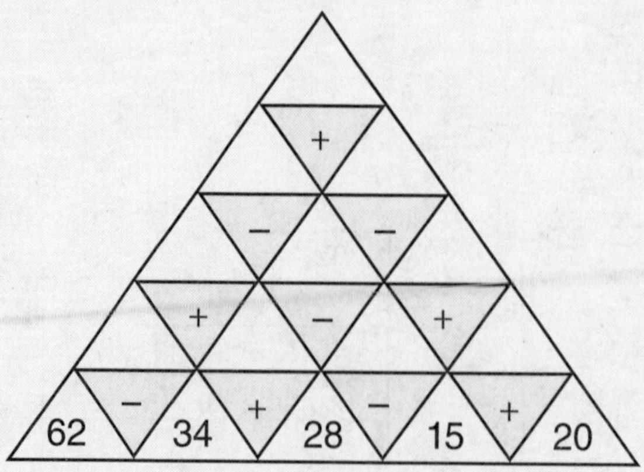

15 Data Handling

Quick Facts

1. Information can be collected by asking people.
2. Information can be presented in tables using pictures.
3. Information can be gathered by reading tables having pictures.

WORKSHEET

How Many?

1. Write the names of your favourite people.

....

Answer the following.

a. How many names have you written? ....

b. How many names begin with the same letter? ....

c. How many names have 4 letters? ....

d. How many names end with the same letter? ....

2. Count and write the number of vegetables that a vegetable seller in your locality has.

Vegetable	i. 	ii. 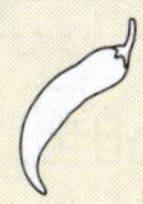	iii. 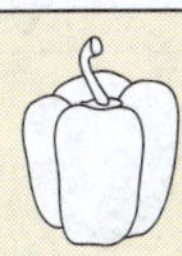	iv. 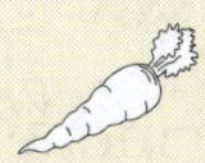	v. 
Number of vegetables					

3. Look at the different vehicles on a road near your house for about 5 minutes on a Sunday. Count them and complete the table.

Vehicle	Number of vehicles
 Bus	
 Auto	
 Car	
 Bike	
 Cycle	

Fill in the blanks.

a. I saw .... autos.

b. Cars  were .... when compared to buses .

c. I saw a total of .... bikes .

d. Very less people used ....

Value-based Question

Arun decided to donate the money in his piggy bank to an orphanage on his birthday.

What value do you learn from him? ....

Now count the money in your piggy bank and tabulate it as given in the table.

Money	Number of coins/notes
1 rupee coin	
2 rupee coin	
5 rupee coin/note	
₹10	
₹20	

How much money do you have altogether? ....

Challenge

How many toys?

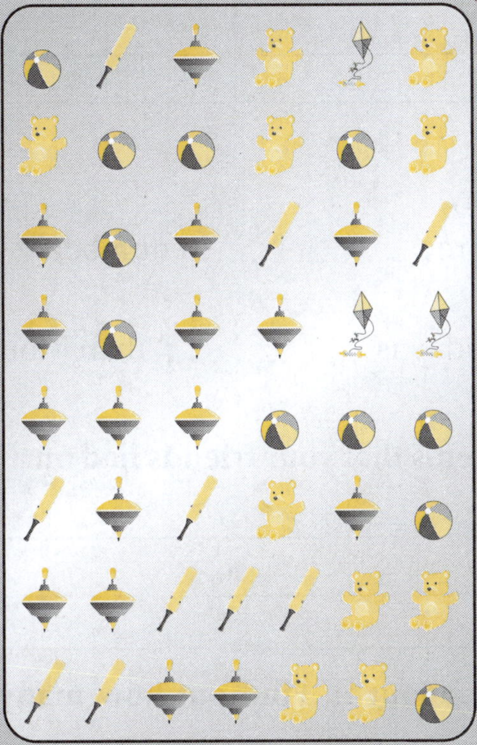

Toy	How many?

Answer the questions.

a. How many  are there? ....

b. Are  less than  ? ....

c.  are .... (more/less) than  .

Revision Worksheet

1. Ask your classmates about their favourite day in the week. Count and write down.

Day							
Number of classmates							

Fill in the blanks.

a. .... classmates like Thursday.

b. Most of them like .... and are .... in number.

c. .... is liked the least.

d. The number of classmates who like Friday is .... than Monday.

e. I like ....

2. Count and write the different breakfast items that your friends had on a particular day.

Breakfast					
Number of friends					

3. Ask your classmates to draw a leaf and colour it. Find out how many students coloured the leaf using the same colours.

Look and write down.

Colour	◯	◯	◯	◯
Number of students				

Colour the circles accordingly.

a. Most of them have coloured the leaf ◯.

b. Students who have coloured the leaf ◯ are less than those who have coloured it ◯.

c. Students who have coloured the leaf ◯ are more than those who have coloured it ◯.

d. ◯ is used the least by the students to colour the leaf.

Reasoning Worksheets

WORKSHEET 1

1. **Which among the following does not belong to the group?**

   a. 
   b. 
   c. 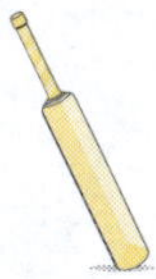
   d. 

2. **How many more designs must be drawn to make it a group of 100 designs?**

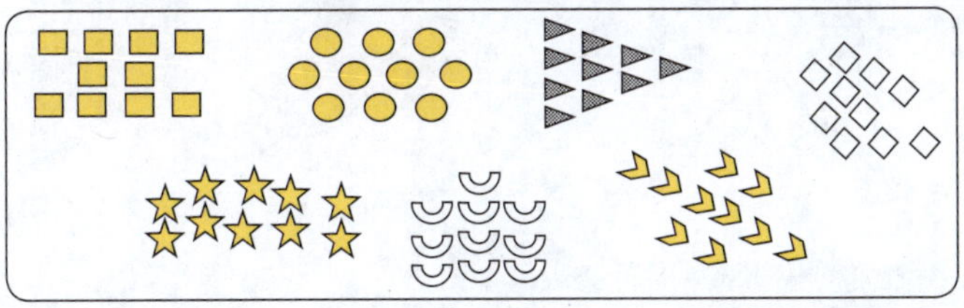

   a. 30 b. 10 c. 40 d. 20

3. **Observe the picture shown and tick on the correct sentence.**

   a. Manu's weight is more than that of Anil
   b. Manu's weight is less than that of Anil
   c. Weights of Anil and Manu are the same
   d. None of the above

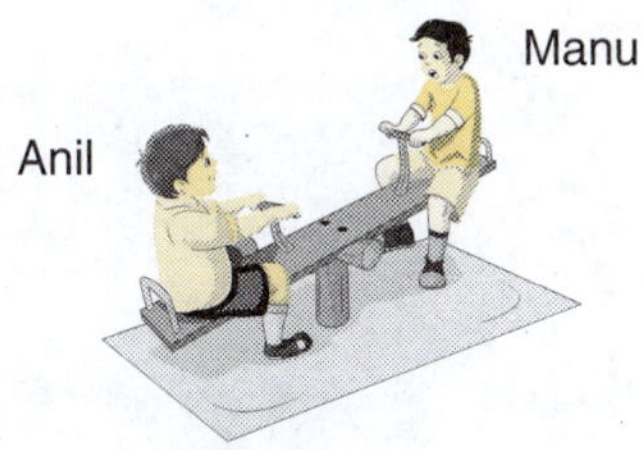

4. **Which shape comes next in the given pattern?**

   a. 
   b. 
   c. 
   d. 

5. **How many  are there in the picture given below?**

a. 7 b. 8 c. 10 d. 11

6. **Which among the following containers holds more water?**

a. 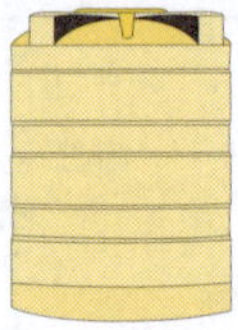 b.  c. 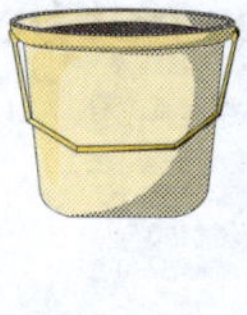 d. 

7. **Identify the combination that does not belong to the group.**

a. 

b. 

c. 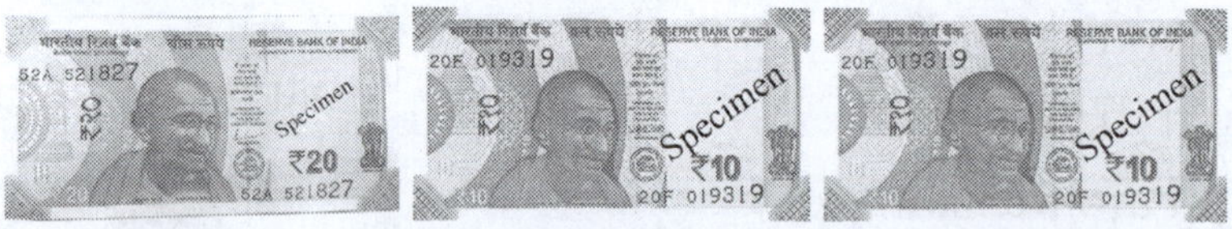

d. 

8. **Which among the following is different from the others?**

a. 48 + 32 b. 75 – 10 c. 36 + 24 d. 85 – 35

9. **Pick the odd one out.**

a. August b. June c. April d September

10. **Identify the false statement.**

a 34 – 16 = 18 b. 42 = 15 + 27 c. 17 = 33 – 14 d. 24 + 26 = 50

WORKSHEET 2

1. Pick the one that is not a member of the group.

a.  b.  c.  d. 

2. How many pairs of objects are there in the picture shown?

a. 14 b. 10 c. 7 d. 5

3. Which object is the lightest among the following?

a.  b.  c.  d. 

4. What comes next in the pattern?

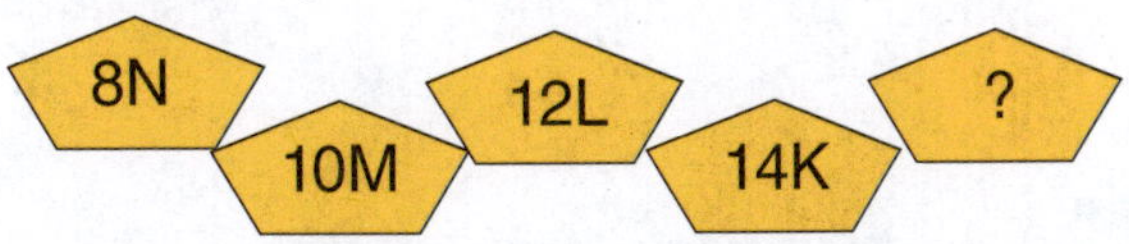

a. 16J b. 18J c. 18I d. 16I

5. Which among the following is the footprint of an elephant?

a.  b.  c. 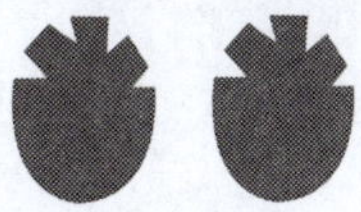 d. 

6. Which among the following holds less liquid?

a.  b.  c.  d. 

7. **Pick the odd one out.**

a. 58 = 53 + 5 b. 40 + 3 = 43 c. 63 = 60 + 3 d. 83 = 80 + 3

8. **Which fruit does not belong to the group?**

a. ORANGE b. MANGO c. GRAPES d. BANANA

9. **Which alphabet is different among the following alphabets?**

a. A b. Y c. H d. P

10. **Vani had ₹98 with her. How much money is left with her after purchasing the following items?**

a. 

b. 

c. 

d. 

ANSWERS

CHAPTER 1

Worksheet 1 (Long and Round objects)

1. Sample answer:
   a. Chalk piece b. Orange c. Calculator d. Pen e. Water bottle
2. a. Coin, Capsicum, Balloon
   b. Clock, glass, scissors
3. a. Sample answer: Coin, ball, apple, pumpkin, orange
   b. Ruler, bat, book, pen, pencil

Worksheet 2 (Things that either slide or roll or both)

1. Rolling objects: 

   Sliding objects: 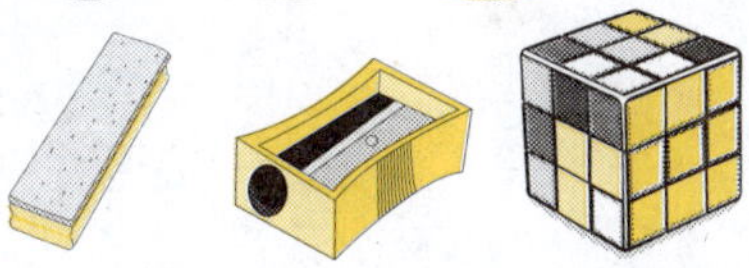

2. 

3. Sample answer:
   Objects that only roll: Orange, lemon, ball, apple
   Objects that only slide: Duster, ruler, book, plate
   Objects that both roll and slide: Coin, marker pen, wheel, bowl

Revision Worksheet

1. a. long b. roll c. round
   d. roll, slide e. roll, slide
2. a. Tick onion, cross-out matchbox
   b. Tick lemon, cross-out suitcase
   c. Tick cucumber, cross-out cube
   d. Tick tyre, cross-out book
   e. Tick apple, cross-out mobile phone

Challenge

Square 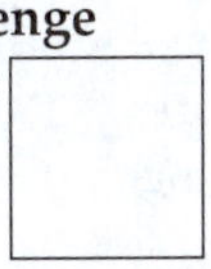

Value-based Question

Tyres are round in shape so that they can roll over a surface smoothly and easily. This will help a vehicle to move faster on a road. If they were square or triangle in shape, they can slide but that will require greater energy. The vehicle will then be forced to move smoothly and it will be difficult for it to move on the road.

Puzzle Corner

L	K	Q	T	M	E	S	N	N	E
L	J	L	L	C	P	T	X	X	G
R	I	W	T	X	O	I	T	P	N
N	O	C	M	A	R	C	X	Z	A
B	W	B	N	O	X	K	N	D	R
U	J	W	E	E	T	D	D	U	O
N	T	G	Q	N	P	A	L	M	F
H	W	R	V	Y	I	E	M	T	T
K	Z	K	I	Q	R	O	R	O	X
L	A	J	N	I	R	B	C	K	T

Things that are round: BOWL, COIN, ORANGE, TOMATO
Things that are long: BRINJAL, PENCIL, RULER, ROPE, STICK

Mental Maths

Sample answer:
a. Scale b. Bangle c. Tomato
d. Notebook e. Wheel

CHAPTER 2

Worksheet 1 (Ordinal Numbers)

1. a. iii b. iv c. v d. i e. ii
2. a. FIRST b. THIRD c. FOURTH d. SEVENTH e. TENTH
3. a. Third b. First c. Tenth d. Sixth e. Eleventh
4. 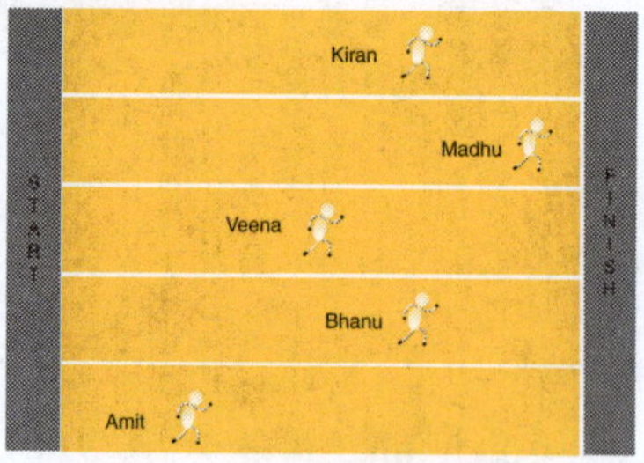

Worksheet 2 (Counting in Groups)

1. a. iii b. iv c. v d. i e. ii
2. a. 18 b. 8 c. 30 d. 28 e. 9
3. a. 20 b. 3 c. 3 d. 20 e. 4
4. 1 group of 24 or 2 groups of 12 or 3 groups of 8 or 4 groups of 6 or 6 groups of 4 or 8 groups of 3 or 12 groups of 2 or 24 groups of 1

Worksheet 3 (Comparison of Numbers)

1. a. 67, 68 b. 21, 24 c. 36, 37 d. 88, 89 e. 52, 53
2. a. > b. < c. = d. < e. <
3. a. 6, 38, 43, 52, 75 b. 8, 27, 38, 80, 98 c. 34, 46, 57, 75, 76
   d. 47, 49, 51, 53, 64 e. 39, 46, 51, 73, 82
4. a. 85, 67, 45, 38, 7 b. 91, 84, 65, 43, 28 c. 87, 56, 48, 37, 12
   d. 72, 68, 63, 43, 19 e. 90, 85, 67, 42, 13

Revision Worksheet

1. a. 12 b. 20 c. 30 d. 24 e. 40
2. a. True b. False c. False d. True e. False
3. a. third b. seventh c. ninth d. second e. fifth
4. a. Descending b. Ascending c. Descending d. Descending e. Ascending

Challenge

Possible 2 digit numbers are 75, 73, 70, 50, 53, 57, 30, 35, 37
Descending order: 75, 73, 70, 57, 53, 50, 37, 35, 30

Puzzle Corner

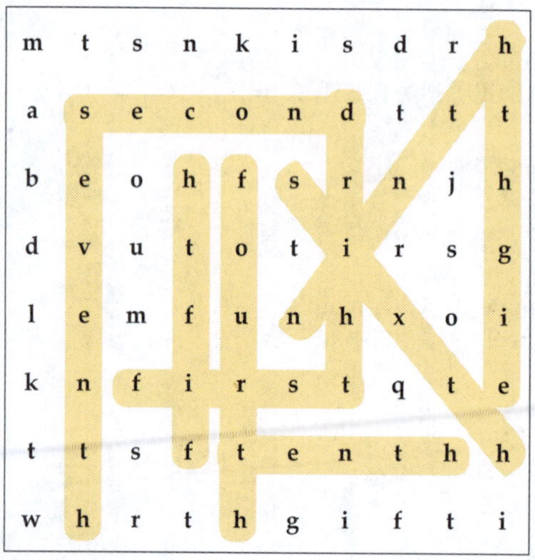

first-1 second-2 third-3 fourth-4
fifth-5 sixth-6 seventh-7 eighth-8
ninth-9 tenth-10

CHAPTER 3

Worksheet (Heavier to Lighter)

1. a. i. ✓ ii. × b. i. ✓ ii. × c. i. × ii. ✓ d. i. × ii. ✓
2. a. i. L ii. H b. i. L ii. H c. i. H ii. L d. i. H ii. L
3. a. heavier b. lighter c. heavier d. lighter e. lighter
4. a. i b. ii c. iii d. ii e. i

Revision Worksheet

1. 

2. a. i. L ii. H b. i. H ii. L c. i. L ii. H d. i. H ii. L
3. a. iv b. v c. i d. ii e. iii

Challenge

Spike is the heaviest and Jerry is the lightest.

Puzzle Corner

CHAPTER 4

Worksheet (Counting in Tens)

1. a. 24 b. 43 c. 31 d. 57 e. 70
2. a. 16 b. 4 c. 70 d. 5 e. 89
3. a. 
   b. 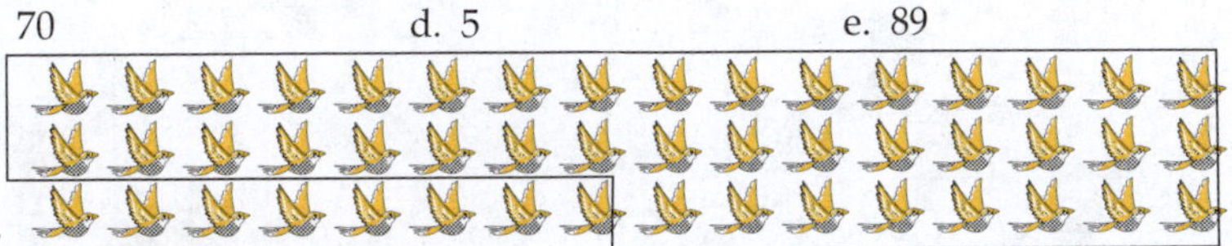
   c. 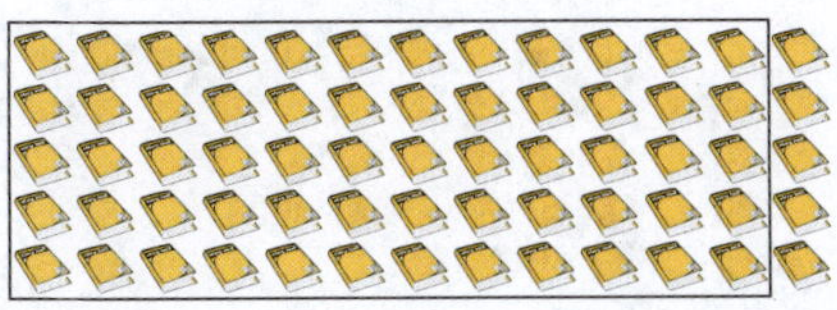
   d. 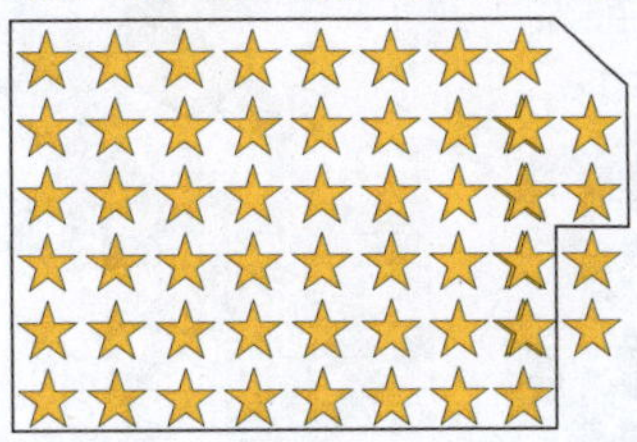
   e. 
4. a. 7 b. 76 c. 3 d. 30, 32 e. 76 – 32 = 44

Challenge

10 + 40 = 50, 20 + 30 = 50, 40 + 10 = 50, 30 + 20 = 50

Revision Worksheet

1. a. iv b. i c. v d. ii e. iii

2. Sample answer: 

3. a. 8 b. 8 c. 88 d. 80 + 8 = 88

Value-based Question

Vani distributed 100 notebooks. The value we learn is helping the needy people. It shows being concerned about spreading education among the people.

Puzzle Corner

Mental Maths

a. 20 + 5 b. 90 c. 2 d. 60 e. 10

CHAPTER 5

Worksheet 1 (Shape patterns)

1. a.  b. 

 c. 

 d. 

2. a.  b.  c. 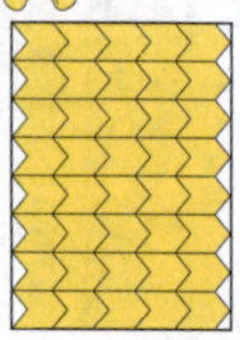 d. 

3. a.  b. 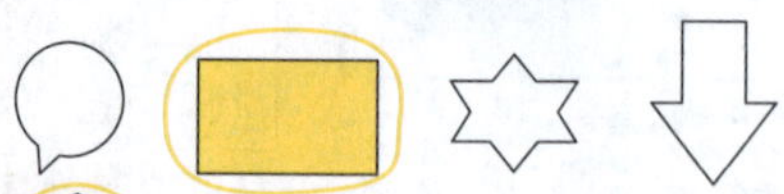

 c. 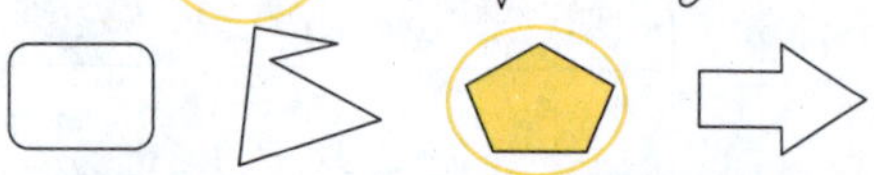 d. 

4. a.  b.  c. 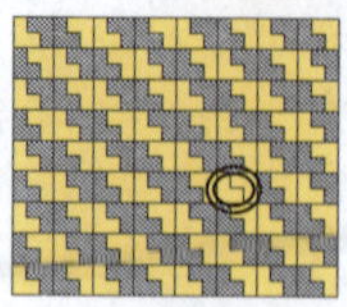 d. No mistake

5. Sample patterns: 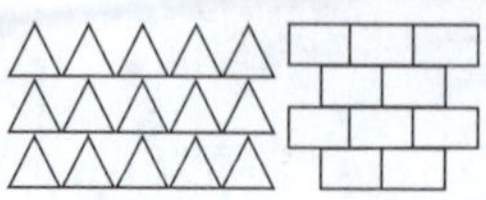

Worksheet 2 (Number patterns)

1. a. Add 5 to the previous number b. Add 2 to the previous number
   c. Subtract 10 from the previous number d. Multiply the previous number by 2
   e. Add 6 to the previous number
2. a. FG b. 123456 c. Left d. F6 e. FFF
3. a. 22 b. 99 c. 7 d. CX e. NM
4. a. 20, 30, 40, 50, 60 b. 85, 80, 75, 70, 65 c. 6, 9, 12, 15, 18
   d. ZZ, YY, XX, WW, VV e. A1A, B2B, C3C, D4D, E5E
5. Sample patterns: AA1, BB2, CC3, DD4 or Aa, Bb, Cc, Dd

Revision Worksheet

1. a. Top and bottom from right side of the circle is coloured alternatively.
   b. Half apple is increased each time.
   c. Write the first letter of the English alphabet twice and number 1; continue the pattern by writing the next letter twice and number 2.
   d. Number 4 is added to the previous number to get the next number.
2. a.  b. 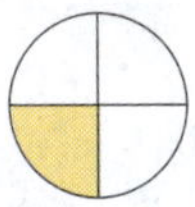 c. 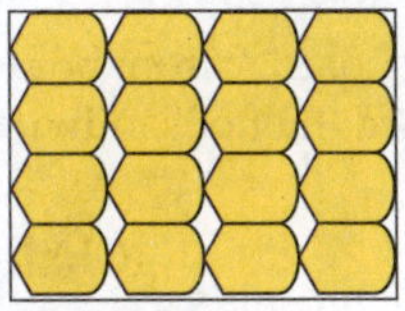 d. 49
3. a. 66 b. 32 c. AAA3 d. 10001
4. a. 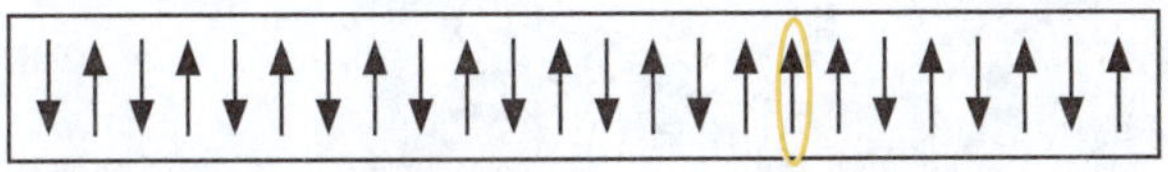 b. 
   c. 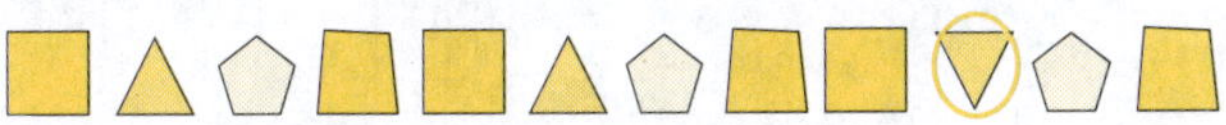
   d. e. ABC, BCD, CDE, DEF, EFE
5. a. 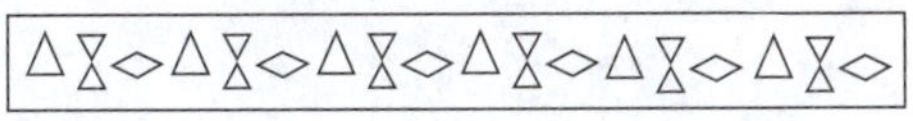 b. 
   c. 5, 8, 11, 14, 17 d. 1A1, 2B2, 3C3, 4D4, 5E5

Challenge

1. 
2. 
3. 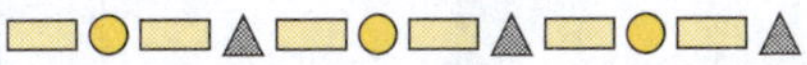
4. 
5. 
6. 
7. 
8. 

Puzzle Corner

17	45	23	13	23	43	53	63
23	36	41	46	51	56	61	14
19	44	33	53	55	58	63	68
33	52	51	52	54	55	56	57
13	60	53	49	58	63	78	88
18	68	43	50	57	64	57	50
79	76	67	64	61	58	55	52

CHAPTER 6

Worksheet (Footprints and Shapes)

1. a. ii b. iii c. iv d. i
2. a. Square b. Circle c. Triangle d. Rectangle

3. 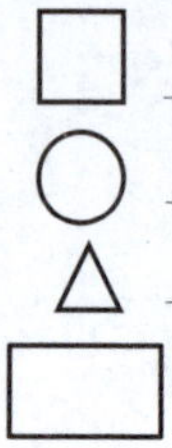

Square	2
Circle	13
Triangle	4
Rectangle	3

4.

Object	Trace below	Name of the shape
a. Eraser		Rectangle
b. Chessboard		Square
c. Coin		Circle

Revision Worksheet

1. Sample answer:
   a. coin, bangle b. handkerchief, chessboard, carom board
   c. biscuit, mat, table d. Pizza piece, traffic signboard, half of sandwich
2. a. ii b. i c. iv d. iii
   Sample answer:

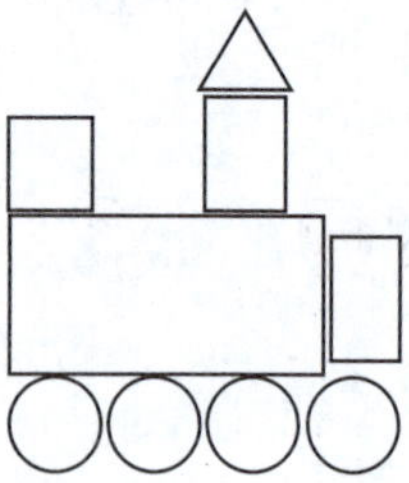

Train engine

Challenge

Puzzle Corner

My hat has:

10 circles, 1 rectangle(s), 1 square, 0 triangles

My face has:

5 circles, 0 rectangles, 0 squares, 1 triangle

My neck is a rectangle. **My arms are** rectangles. **My body is a** triangle. **My legs are** rectangles.

Each of my foot look like a rectangle and is split into 2 triangles.

CHAPTER 7

Worksheet 1 (Estimation of Capacity)

1. a. A jug b. ₹100 c. 12 short glasses d. 6 short glasses e. ₹80
2. Sample answer:
   a. 5 b. 10 c. 7 d. 20 e. 30
3. Sample answer:

Activity	Amount of Water
Taking bath	8 buckets
Washing clothes	30 buckets
Cooking and cleaning utensils	20 buckets
Drinking	10 jugs

Worksheet 2 (Comparison of Capacities)

1. a. i b. ii c. ii d. i e. ii
2. a. ii b. iii c. i
3. Sample answer:

A milk cup

4. a. The container is not completely filled. b. The container is completely filled and water falls out.
   c. The glass is completely filled and water falls out. d. The bucket is not completely filled.

Challenges

c.

Puzzle Corner

Mental Maths

a. Spoon b. Around 8 c. Water tank d. Sample answer: 5 buckets

Revision Worksheet

1. Sample answer: A spoon, a cup, a bowl
2. Sample answer: A jug, a water bottle, a bucket
3. a. ₹75 b. ₹90 c. 3 d. Sample answer: 2 glasses

Value-based Question

She poured the buttermilk from the buckets into a jug and then from the jug into glasses.
Value: Respect elders and treat them well.

CHAPTER 8

Worksheet 1 (Tens and Ones)

1. a. v b. iv c. i d. ii e. iii
2. a. 30 b. 9 c. 70 d. 5 e. 90
3. a. 43 b. 69 c. 96 d. 38 e. 90
4.

	Number
b. 	23
c. 	48
d. 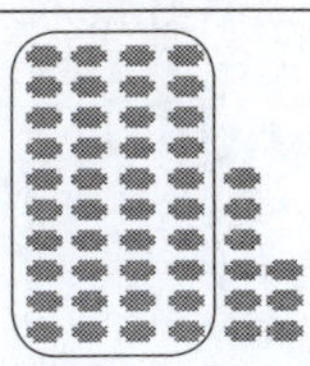	49

Worksheet 2 (Real Life Applications)

1. Sample answer:
   a. ₹2 + ₹2 or ₹1 + ₹1 + ₹1 + ₹1
   b. ₹10 + ₹5 or ₹5 + ₹5 + ₹5
   c. ₹20 + ₹3 or ₹10 + ₹10 + ₹2 + ₹1
   d. ₹20 + ₹10 or ₹10 + ₹10 + ₹10
   e. ₹20 + ₹20 + ₹5 + ₹2 + ₹1 or ₹20 + ₹10 + ₹10 + ₹5 + ₹2 + ₹1
2. a. 3 b. 30 c. 1 d. 99 e. 9
3. a. 14 b. 3 c. 7 d. 4 e. 7

Revision Worksheet

1. b. 5 tens + 5 ones c. 2 tens + 7 ones d. 8 tens + 1 ones e. 9 tens + 9 ones
2. a. × b. ✓ c. × d. ✓ e. ✓
3. a. 32 b. 40 c. 72 points:  d. 7, 2

Challenge

4 notes of ₹10 = ₹40
2 notes of ₹20 = ₹40
8 notes of ₹5 = ₹40
8 coins of ₹5 = ₹40
20 coins of ₹2 = ₹40
40 coins of ₹1 = ₹40
2 notes of ₹10 + 1 note of ₹20 = ₹40
2 coins of ₹5 + 1 note of ₹20 + 1 note of ₹10 = ₹40

CHAPTER 9

Worksheet 1 (Days of the week)

1. a. Wednesday b. Monday c. Thursday d. Sunday e. Friday
2. Sample answer:

Period	Art and craft	English	GK	Music	Pay time
Day of the Week	Wednesday	Friday	Tuesday	Thursday	Monday

3. a. F b. T c. F d. T e. F

Worksheet 2 (Months in a year)

1. Sample answer:

Name of the friend	Birthday month
Navya	March
Dhruv	July
Vani	May
Manu	August

2. a. March b. October c. December
   d. July e. February
3. a. April, June, September, November
   b. January, March, May, July, August, October, December
   c. 61 days d. 4 e. 28 days
4. a. January b. Depends on the year c. December d. October
   e. Depends on the year
5. Sample answer:

My favourite	Name of the month when it grows
Flower: Lily	February
Fruit: Mango	May
Vegetable: Capsicum	July

Challenge

1. September 2. Sunday

Revision Worksheet

1. a. Thursday b. December c. May d. October e. 30
2. a. iii b. iv c. v d. ii e. i
3. a. Friday b. Monday c. Saturday d. Thursday e. Sunday

Value-based Question

Dhruv celebrated his birthday on Wednesday of May.
Value: To serve the needy is a good habit.

Puzzle Corner

F	B	J	A	N	U	A	R	Y
Y	E	Y	A	D	S	E	U	T
A	J	B	B	L	B	Q	Y	X
D	U	N	R	M	M	A	Y	Y
I	N	B	E	U	D	J	B	T
R	E	C	T	N	A	R	W	M
F	E	T	U	W	R	R	T	D
D	N	S	G	G	Z	Z	Y	L

1. February **2.** May **3.** Tuesday **4.** Friday
5. June **6.** Sunday **7.** January **8.** December

CHAPTER 10

Worksheet 1 (Addition of 2 numbers)

1. a. 61 b. 35 c. 64 d. 93 e. 91
2. a. 5 b. 7 c. 6 d. 4 e. 9
3. a. 15 + 18 b. 20 + 23 c. 25 + 29 d. 30 + 35 e. 29 + 45
4. a. iii b. iv c. v d. i e. ii

Worksheet 2 (Addition of 3 or more numbers)

1. a. 24 b. 16 c. 26 d. 22 e. 25
2. Sample answer:
   a. 10 + 5 + 3 b. 10 + 10 + 7 c. 10 + 10 + 10 + 3
   d. 5 + 5 + 5 + 5 e. 4 + 4 + 4 + 3
3. a. 12 + 13 + 4 b. 10 + 3 + 2 c. 40 + 10 + 2 d. 30 + 5 + 3 e. 5 + 3 + 2
4. a. T b. T c. F d. T e. F

Challenge

Puzzle Corner

				17
2	8	5	7	22
6	4	0	6	16
0	9	7	3	19
1	2	3	1	7
9	23	15	17	14

Revision Worksheet

1. Sample answer:
   a. 88 = 80 + 8 or 85 + 3 b. 56 = 50 + 6 or 55 + 1 c. 62 = 60 + 2 or 55 + 7
   d. 17 = 10 + 7 or 15 + 2 e. 33 = 30 + 3 or 25 + 8
2. a. 5 b. 8 c. 9 d. 3 e. 45
3. a. ii b. iv c. i d. v e. iii
4. a. F b. T c. F d. T e. T

CHAPTER 11

Worksheet 1 (Types of Lines)

1. a. Straight line b. Straight line c. Curved line d. Curved line
   e. Straight line f. Curved line
2. a. 3, 4 b. 4, 2 c. 3, 3
3. Straight lines: A, E, F, H, I, K, L, M, N, T, V, W, X, Y, Z
   Curved lines: C, O, S
   Both straight and curved lines: B, D, G, J, P, Q, R, U

Worksheet 2 (Shapes and Pictures with Lines)

1. a. ii b. ii c. ii d. i e. ii
2.

Figure	Standing lines	Sleeping lines	Slanting lines	Curved lines
a	4	5	2	0
b	3	0	2	3
c	2	3	2	0
d	0	5	2	6

3.

Only straight lines	Only curved lines	Both straight and curved lines
	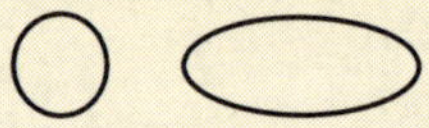	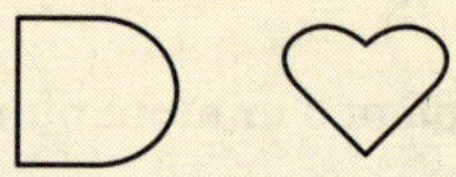

Revision Worksheet

1. a. 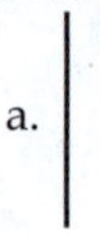 b.  c.  d. 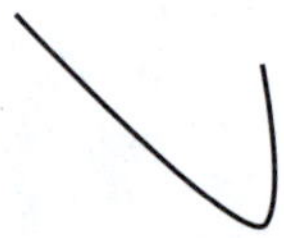
2. b. Only curved lines c. Only straight lines d. Both straight and curved lines
3. a. ✓ b. × c. ✓ d. ×
4.

Straight lines	Curved lines	Both straight and curved lines
N, F, L, A, K, E 1, 4, 7	C, O, S, 3, 8, 9, 0	R, 2, 5

Challenge

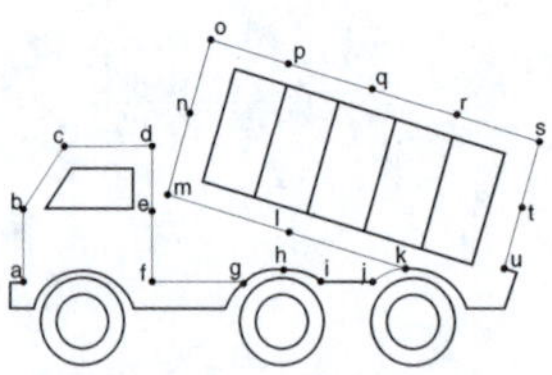

A total of 45 straight lines and 12 curved lines are used to draw the picture.

Puzzle Corner

Sample answer

CHAPTER 12

Worksheet 1 (Addition)

1. a. 40 b. 60 c. 60 d. 85 e. 83
2. a. (40), 50 b. (61), 70
3. a. 15 and 25 b. 17 and 18 c. 18 and 26 d. 55 and 39

Worksheet 2 (Subtraction)

1. a. 18 b. 6 c. 22 d. 38 e. 49
2. a. 17 b. 18 c. 32 d. 28 e. 34
3. a. 18 b. 19 c. 37 d. 49 e. 30 f. 63

Worksheet 3 (Word Problems)

1. 45 students 2. ₹82 3. 64 beads 4. 31 seats 5. 15 litres

Challenge

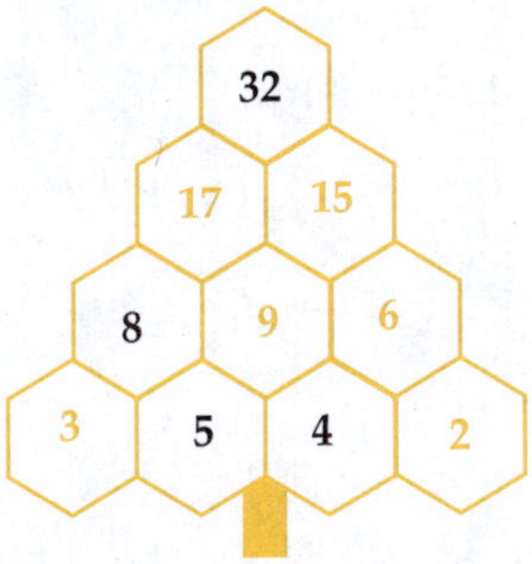

Revision Worksheet

1. a. iii b. i c. v d. iv e. ii
2. a. 91 b. 23 c. 60 d. 85 e. 29
3. a. 72 b. 21 c. 95 d. 32 e. 28
4. a. 35 mangoes b. 94 stickers c. 79 points d. ₹14
   e. 36 red-coloured papers

Value-based Question

a. 13 + [2] = 11 + 4 b. 9 + 8 = [3] + 14
c. 10 + [5] = 12 + 3 d. 15 + 2 = [10] + 7
e. [3] + 14 = 12 + 5 f. [5] + 12 = 9 + 8
g. 2 + 14 = [3] + 13 h. [6] + 12 = 14 + 4

Mental Maths

1. 40 2. 71 3. 61
4. 80 5. 60
6. a. 17 b. 44 c. 13 d. 17 e. 42

Puzzle Corner

8	+	5	+	10	=	23
+		+		+		+
9	+	7	+	10	=	26
+		+		+		+
10	+	10	+	9	=	29
=		=		=		=
27	+	22	+	29	=	78

CHAPTER 13

Worksheet (Finger length, handspan, footspan and pace)

1. a. Handspan b. Cubit c. Footspan d. Pace
2. Finger length, Handspan, Handspan, Finger length
3. a. 9 b. 4 c. 7 d. 6

Revision Worksheet

1. a. 7 handspan b. 4 handspan c. 3 cubits
2. a. F b. F c. T d. T e. T
3. Sample answer: Crayons, Eraser, Sharpener, Safety pin

Challenge

Sample answer:

1. 1 cubit = 3 handspans and 8 fingers
2. 1 pace = 4 footspans

CHAPTER 14

Worksheet 1 (Addition)

1. a. 41 b. 75 c. 81 d. 91 e. 82
2. a. ii b. v c. i d. iii e. iv
3. Sample answer:
   a. 20 + 34 b. 40 + 43 c. 15 + 30 d. 30 + 46 e. 21 + 70
4. a. 63 garlands b. 99 marks c. 83 vehicles d. ₹80 e. 90 cups

Worksheet 2 (Subtraction)

1. a. 14 b. 27 c. 36 d. 42 e. 39
2. a. 16 b. 28 c. 73 d. 57 e. 34
3. Sample answer:
4. a. 40 – 16 b. 60 – 19 c. 70 – 35 d. 90 – 28 e. 85 – 32
5. a. 27 pencils b. 18 chocolates c. Manu, 16 balloons more d. 56 years e. ₹44

Worksheet 3 (Real Life Applications)

1. a. ₹91 b. Yes c. A train and a top or an aeroplane, bat and ball
   d. ₹17 e. ₹33
2. a. 34 children b. Least liked: watermelon; most liked: mango c. 19 children d. 26 children

Revision Worksheet

1. a. ii b. iii c. i d. v e. iv
2. a, d and e
3. a. 37 cookies b. 81 stamps c. 33 d. ₹92 e. ₹64

Challenge

37	+	26	=	63
+		+		
45	-	18	=	27
=		=		
82	-	44	=	38

Puzzle Corner

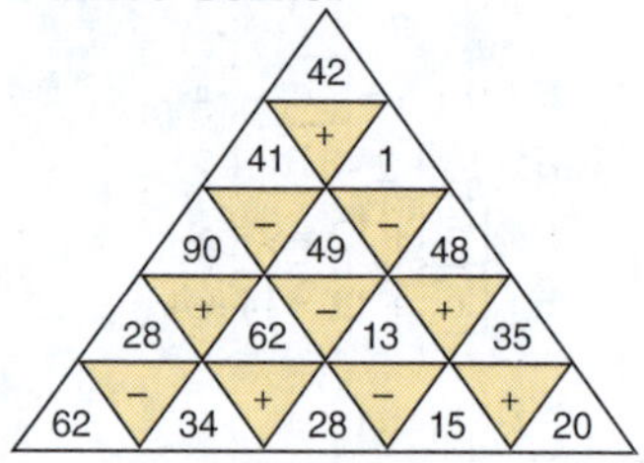

Value-based Question

₹83. Helping the needy is a good habit.

Mental Maths

1. 84 2. 69 3. 60 4. 80
5. 38 6. 90 7. 80 8. 17

CHAPTER 15

Worksheet (How Many?)

Sample answers are given.

1. Anand, Sanvi, Dhruv, Vani, Manu, Shyam, Anil, Vidya, Padma
   a. 9 b. 6 c. 3 d. 4
2. i. 40 ii. 60 iii. 25 iv. 18 v. 10
3. Bus: 5, Auto: 12, Car: 10, Bike: 25, Cycle: 3
   a. 12 b. more c. 25 d. cycles

Value-based Question

Serving the needy people is a good habit.

Challenge

Bat: 10 Top: 16 Soft toy: 10 Ball: 11 Kite: 3; a. 11 b. No c. more

Revision Worksheet

Work along with your classmates and collect data to complete the worksheet.

REASONING WORKSHEETS

Worksheet 1

1. d 2. a 3. b 4. c 5. a
6. a 7. d 8. b 9. a 10. c

Worksheet 2

1. a 2. c 3. d 4. a 5. c
6. b 7. a 8. b 9. d 10. a